AF581740

3rd XoveTIC Conference

3rd XoveTIC Conference

Special Issue Editors

Joaquim de Moura
Alejandro Puente-Castro
Javier Pereira Loureiro
Manuel G. Penedo

MDPI • Basel • Beijing • Wuhan • Barcelona • Belgrade

Editors
Joaquim de Moura
University of A Coruña
Spain

Alejandro Puente-Castro
University of A Coruña
Spain

Javier Pereira Loureiro
University of A Coruña
Spain

Manuel G. Penedo
University of A Coruña
Spain

Editorial Office
MDPI
St. Alban-Anlage 66
4052 Basel, Switzerland

This is a reprint of articles from the Special Issue published online in the open access journal *Proceedings* (ISSN 2504-3900) (available at: https://www.mdpi.com/2504-3900/54/1).

For citation purposes, cite each article independently as indicated on the article page online and as indicated below:

LastName, A.A.; LastName, B.B.; LastName, C.C. Article Title. *Journal Name* **Year**, *Volume Number*, Page Range.

ISBN 978-3-0365-0856-6 (Hbk)
ISBN 978-3-0365-0857-3 (PDF)

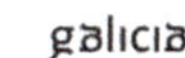
galicia

citic

UNIVERSIDADE DA CORUÑA

Xacobeo 2021

Unión Europea

XUNTA
DE GALICIA

Contents

List of Contributors xi

About the Special Issue Editors xiii

Preface to "Proceedings 3rd XoveTIC Conference" xv

Arturo Silvelo, Daniel Garabato, Raúl Santoveña and Carlos Dafonte
A First Approach to Authentication Based on Artificial Intelligence for Touch-Screen Devices
Reprinted from: *Proceedings* **2020**, *54*, 1, doi:10.3390/proceedings2020054001 1

Luis M. Hervella-Nieto, Andrés Prieto and Sara Recondo
Computation of Resonance Modes in Open Cavities with Perfectly Matched Layers
Reprinted from: *Proceedings* **2020**, *54*, 2, doi:10.3390/proceedings2020054002 4

Celtia Meizoso Lamas, Francisco Bellas and Berta Guijarro-Berdiñas
SARDAM: Service Assistant Robot for Daily Activity Monitoring
Reprinted from: *Proceedings* **2020**, *54*, 3, doi:10.3390/proceedings2020054003 6

Aida Vidal-Balea, Oscar Blanco-Novoa, Paula Fraga-Lamas, Miguel Vilar-Montesinos and Tiago M. Fernández-Caramés
A Collaborative Augmented Reality Application for Training and Assistance during Shipbuilding Assembly Processes
Reprinted from: *Proceedings* **2020**, *54*, 4, doi:10.3390/proceedings2020054004 9

Martiño Rivera Dourado, Marcos Gestal and José M. Vázquez-Naya
Implementing a Web Application for W3C WebAuthn Protocol Testing
Reprinted from: *Proceedings* **2020**, *54*, 5, doi:10.3390/proceedings2020054005 12

Felpeto-Guerrero Abraham, Vázquez-Sánchez Rubén, Chao-Fernández Rocío and Chao-Fernández Aurelio
Use of Arduino Microcontroller in Education: Creation of "The Musical Stairs"
Reprinted from: *Proceedings* **2020**, *54*, 6, doi:10.3390/proceedings2020054006 15

Iñigo López-Riobóo Botana, Carlos Eiras-Franco and Amparo Alonso-Betanzos
Regression Tree Based Explanation for Anomaly Detection Algorithm
Reprinted from: *Proceedings* **2020**, *54*, 7, doi:10.3390/proceedings2020054007 18

Celtia Meizoso Lamas, Francisco Bellas and Berta Guijarro-Berdiñas
SARDAM: Service Assistant Robot for Daily Activity Monitoring
Reprinted from: *Proceedings* **2020**, *54*, 8, doi:10.3390/proceedings2020054008 22

Shaghayegh Ravaei, Juan M. Alonso-Martinez, Alberto Jimenez-Zayas and Francisco Sendra-Portero
Open AccessProceedingsPeer-ReviewedReflections about Learning Radiology inside the Multi-User Immersive Environment Second Life® during Confinement by Covid-19
Reprinted from: *Proceedings* **2020**, *54*, 9, doi:10.3390/proceedings2020054009 26

Luís Fernández-Núñez, Darío Penas, Jorge Viteri, Carlos Gómez-Rodríguez and Jesús Vilares
Developing Open-Source Roguelike Games for Visually-Impaired Players by Using Low-Complexity NLP Techniques
Reprinted from: *Proceedings* **2020**, *54*, 10, doi:10.3390/proceedings2020054010 30

Eva Blanco-Mallo, Beatriz Remeseiro, Verónica Bolón-Canedo and Amparo Alonso-Betanzos
On the Effectiveness of Convolutional Autoencoders on Image-Based Personalized Recommender Systems
Reprinted from: *Proceedings* **2020**, *54*, 11, doi:10.3390/proceedings2020054011 . . . . **34**

Andrés Molares, Enrique Fernandez-Blanco and Daniel Rivero
Application of Artificial Neural Networks for the Monitoring of Episodes of High Toxicity by DSP in Mussel Production Areas in Galicia
Reprinted from: *Proceedings* **2020**, *54*, 12, doi:10.3390/proceedings2020054012 . . . . **37**

María González and Hiram Varela
Numerical Simulation of a Nonlinear Problem Arising in Heat Transfer and Magnetostatics
Reprinted from: *Proceedings* **2020**, *54*, 13, doi:10.3390/proceedings2020054013 . . . . **41**

Beatriz Salvador, Cornelis W. Oosterlee and Remco van der Meer
European and American Options Valuation by Unsupervised Learning with Artificial Neural Networks
Reprinted from: *Proceedings* **2020**, *54*, 14, doi:10.3390/proceedings2020054014 . . . . **44**

Nereida Rodriguez-Fernandez, Iria Santos, Alvaro Torrente-Patiño and Adrian Carballal
Digital Image Quality Prediction System
Reprinted from: *Proceedings* **2020**, *54*, 15, doi:10.3390/proceedings2020054015 . . . . **47**

Juan M. Alonso-Martinez, Shaghayegh Ravaei, Teodoro Rudolphi-Solero and Francisco Sendra-Portero
Radiology Seminars with Guest Professors in the Virtual Environment Second Life®: Perception of Learners and Teachers
Reprinted from: *Proceedings* **2020**, *54*, 16, doi:10.3390/proceedings2020054016 . . . . **51**

Alberto-Jimenez-Zayas, Shaghayegh Ravaei, Juan M. Alonso-Martinez and Francisco Sendra-Portero
Preliminary Analysis of a Virtual Inter-University Game to Learn Radiology within the Second Life® Environment
Reprinted from: *Proceedings* **2020**, *54*, 17, doi:10.3390/proceedings2020054017 . . . . **55**

Teodoro Rudolphi-Solero, Alberto Jimenez-Zayas and Francisco Sendra-Portero
Introduction to Second Life® As a Virtual Training Environment: Perception of University Teachers
Reprinted from: *Proceedings* **2020**, *54*, 18, doi:10.3390/proceedings2020054018 . . . . **58**

Patricia Concheiro-Moscoso, María del Carmen Miranda-Duro, Carlota Fraga, Cristina Queirós, António José Pereira da Silva Marques and Betania Groba
Design of a System to Implement Occupational Stress Studies Trough Wearables Devices and Assessment Tests
Reprinted from: *Proceedings* **2020**, *54*, 19, doi:10.3390/proceedings2020054019 . . . . **62**

María del Carmen Miranda-Duro, Patricia Concheiro-Moscoso, Javier Lagares Viqueira, Laura Nieto-Riveiro, Nereida Canosa Domínguez and Thais Pousada García
Virtual Reality Game Analysis for People with Functional Diversity: An Inclusive Perspective
Reprinted from: *Proceedings* **2020**, *54*, 20, doi:10.3390/proceedings2020054020 . . . . **66**

Alberto Femenias-Hermida, Cristian R. Munteanu and José M. Vázquez-Naya
Design and Implementation of a Physical Bitcoin Coin
Reprinted from: *Proceedings* **2020**, *54*, 21, doi:10.3390/proceedings2020054021 . . . . **69**

Manuel Seijas Carpente, Bertha Guijarro-Berdiñas, Amparo Alonso-Betanzos, Alejandro Rodríguez-Arias and Adina Dimitru
An Agent-Based Model to Simulate the Spread of a Virus Based on Social Behavior and Containment Measures
Reprinted from: *Proceedings* **2020**, *54*, 22, doi:10.3390/proceedings2020054022 . . . . **72**

Carlos Fernandez-Lozano and Francisco Cedron
Shiny Dashboard for Monitoring the COVID-19 Pandemic in Spain
Reprinted from: *Proceedings* **2020**, *54*, 23, doi:10.3390/proceedings2020054023 . . . . **75**

Iván Froiz-Míguez, Paula Fraga-Lamas and Tiago M. Fernández-Caramés
Decentralized P2P Broker for M2M and IoT Applications
Reprinted from: *Proceedings* **2020**, *54*, 24, doi:10.3390/proceedings2020054024 . . . . **78**

Álvaro S. Hervella, Lucía Ramos, José Rouco, Jorge Novo and Marcos Ortega
Joint Optic Disc and Cup Segmentation Using Self-Supervised Multimodal Reconstruction Pre-Training
Reprinted from: *Proceedings* **2020**, *54*, 25, doi:10.3390/proceedings2020054025 . . . . **81**

Paulo Veloso Gomes, João Donga, António Marques, João Azevedo and Javier Pereira
Analysis and Definition of Data Flows Generated by Bio Stimuli in the Design of Interactive Immersive Environments
Reprinted from: *Proceedings* **2020**, *54*, 26, doi:10.3390/proceedings2020054026 . . . . **84**

Estefanía López Salas, Adrián Xuíz García, Ángel Gómez and Carlos Dafonte
CultUnity3D: A Virtual Spatial Ecosystem for Digital Engagement with Cultural Heritage Sites
Reprinted from: *Proceedings* **2020**, *54*, 27, doi:10.3390/proceedings2020054027 . . . . **87**

Paula Carracedo-Reboredo, Cristian R. Munteanu, Humbert González-Díaz and Carlos Fernandez-Lozano
Web Server and R Library for the Calculation of Markov Chains Molecular Descriptors
Reprinted from: *Proceedings* **2020**, *54*, 28, doi:10.3390/proceedings2020054028 . . . . **91**

Francisco José Martínez-Martínez, Patricia Concheiro-Moscoso, María Del Carmen Miranda-Duro, Francisco Docampo Boedo, Francisco Javier Mejuto Muiño and Betania Groba
Validation of Self-Quantification Xiaomi Band in a Clinical Sleep Unit
Reprinted from: *Proceedings* **2020**, *54*, 29, doi:10.3390/proceedings2020054029 . . . . **94**

Anxo Pérez, Paula Lopez-Otero and Javier Parapar
Designing an Open Source Virtual Assistant
Reprinted from: *Proceedings* **2020**, *54*, 30, doi:10.3390/proceedings2020054030 . . . . **98**

Joaquim de Moura, Lucía Ramos, Plácido L. Vidal, Jorge Novo and Marcos Ortega
Analysis of Separability of COVID-19 and Pneumonia in Chest X-ray Images by Means of Convolutional Neural Networks
Reprinted from: *Proceedings* **2020**, *54*, 31, doi:10.3390/proceedings2020054031 . . . . **101**

Lucía Ramos, Jorge Novo, José Rouco, Stéphanie Romeo, María D. Álvarez and Marcos Ortega
Fully Automatic Retinal Vascular Tortuosity Assessment Integrating Domain-Related Information
Reprinted from: *Proceedings* **2020**, *54*, 32, doi:10.3390/proceedings2020054032 . . . . **104**

Mateo Gende, Marcos Ortega and Manuel G. Penedo
Data Extraction in Insurance Photo-Inspections Using Computer Vision
Reprinted from: *Proceedings* **2020**, *54*, 33, doi:10.3390/proceedings2020054033 107

Sara Alvarez-Gonzalez, Jose Rodriguez-Fernandez, Ana Belen Porto-Pazos, Alejandro Pazos and Francisco Cedron
Probiotic: First Prescriptive Application of Probiotics in Spain
Reprinted from: *Proceedings* **2020**, *54*, 34, doi:10.3390/proceedings2020054034 110

Tiago Gonçalves, Wilson Silva, Maria J. Cardoso and Jaime S. Cardoso
Deep Image Segmentation for Breast Keypoint Detection
Reprinted from: *Proceedings* **2020**, *54*, 35, doi:10.3390/proceedings2020054035 112

Alejandro Rodríguez-Arias, Bertha Guijarro-Berdiñas and Noelia Sánchez-Maroño
An Intelligent and Collaborative Multiagent System in a 3D Environment
Reprinted from: *Proceedings* **2020**, *54*, 36, doi:10.3390/proceedings2020054036 116

Jose Balsa
Comparison of Image Compressions: Analog Transformations
Reprinted from: *Proceedings* **2020**, *54*, 37, doi:10.3390/proceedings2020054037 119

David Novoa-Paradela, Óscar Fontenla-Romero and Bertha Guijarro-Berdiñas
Adaptive Real-Time Method for Anomaly Detection Using Machine Learning
Reprinted from: *Proceedings* **2020**, *54*, 38, doi:10.3390/proceedings2020054038 122

Marc Bernice Angoue Avele, Julio Brégains, Roberto Maneiro, José A. García-Naya and Luis Castedo
Numerical Model of a 28-GHz Frequency Diverse Array Antenna
Reprinted from: *Proceedings* **2020**, *54*, 39, doi:10.3390/proceedings2020054039 125

J. Guzmán Figueira-Domínguez, Verónica Bolón-Canedo and Beatriz Remeseiro
Feature Selection in Big Image Datasets
Reprinted from: *Proceedings* **2020**, *54*, 40, doi:10.3390/proceedings2020054040 128

Alberto Alvarellos and Juan Ramón Rabuñal
Developing an Open-Source, Low-Cost, Radon Monitoring System
Reprinted from: *Proceedings* **2020**, *54*, 41, doi:10.3390/proceedings2020054041 131

João Donga, Paulo Veloso Gomes, António Marques, Javier Pereira and João Azevedo
Application of Adaptive Virtual Environments Through Biofeedback for the Treatment of Phobias
Reprinted from: *Proceedings* **2020**, *54*, 42, doi:10.3390/proceedings2020054042 134

Catarina Sá, Paulo Veloso Gomes, António Marques and António Correia
The Use of Portable EEG Devices in Development of Immersive Virtual Reality Environments for Converting Emotional States into Specific Commands
Reprinted from: *Proceedings* **2020**, *54*, 43, doi:10.3390/proceedings2020054043 137

José Morano, Álvaro S. Hervella, Noelia Barreira, Jorge Novo and José Rouco
Enhancing Retinal Blood Vessel Segmentation through Self-Supervised Pre-Training
Reprinted from: *Proceedings* **2020**, *54*, 44, doi:10.3390/proceedings2020054044 140

Plácido L. Vidal, Joaquim de Moura, Lucía Ramos, Jorge Novo and Marcos Ortega
Study on Relevant Features in COVID-19 PCR Tests
Reprinted from: *Proceedings* **2020**, *54*, 45, doi:10.3390/proceedings2020054045 143

David Moreno Naya, Francisco J. Vazquez-Araujo, Paula M. Castro, Jamile Vivas Costa, Adriana Dapena and Luz González Doniz
Mobile Application for Analysing the Development of Motor Skills in Children
Reprinted from: *Proceedings* **2020**, *54*, 46, doi:10.3390/proceedings2020054046 . . . . **146**

João Azevedo, Paulo Veloso Gomes, João Donga and António Marques
A-Frame as a Tool to Create Artistic Collective Installations in Virtual Reality
Reprinted from: *Proceedings* **2020**, *54*, 47, doi:10.3390/proceedings2020054047 . . . . **150**

Sara Guerreiro-Santalla, Francisco Bellas and Richard J. Duro
Artificial Intelligence in Pre-University Education: What and How to Teach
Reprinted from: *Proceedings* **2020**, *54*, 48, doi:10.3390/proceedings2020054048 . . . . **153**

Mireia Martí Vilas, Pilar Vila Avendaño and José M. Vázquez-Naya
Development of an Open Source Tool and a Multi-Platform for Generation of Forensic Reports
Reprinted from: *Proceedings* **2020**, *54*, 49, doi:10.3390/proceedings2020054049 . . . . **157**

Diego Fernández-Edreira, Jose Liñares-Blanco and Carlos Fernandez-Lozano
Identification of Prevotella, Anaerotruncus and Eubacterium Genera by Machine Learning Analysis of Metagenomic Profiles for Stratification of Patients Affected by Type I Diabetes
Reprinted from: *Proceedings* **2020**, *54*, 50, doi:10.3390/proceedings2020054050 . . . . **160**

Pedro Cabalar, Rodrigo Martín, Brais Muñiz and Gilberto Pérez
aspBEEF: Explaining Predictions Through Optimal Clustering
Reprinted from: *Proceedings* **2020**, *54*, 51, doi:10.3390/proceedings2020054051 . . . . **162**

Iker González-Santamaría, Minia Manteiga, Carlos Dafonte, Arturo Manchado and Ana Ulla
Mining of the Milky Way Star Archive Gaia-DR2. Searching for Binary Stars in Planetary Nebulae
Reprinted from: *Proceedings* **2020**, *54*, 52, doi:10.3390/proceedings2020054052 . . . . **165**

Francisco Laport, Paula M. Castro, Adriana Dapena, Francisco J. Vazquez-Araujo and Daniel Iglesia
Study of Machine Learning Techniques for EEG Eye State Detection
Reprinted from: *Proceedings* **2020**, *54*, 53, doi:10.3390/proceedings2020054053 . . . . **168**

Bieito Beceiro, Jorge González-Domínguez and Juan Touriño
Acceleration of a Feature Selection Algorithm Using High Performance Computing
Reprinted from: *Proceedings* **2020**, *54*, 54, doi:10.3390/proceedings2020054054 . . . . **171**

Rebeca Peláez Suárez, Ricardo Cao Abad and Juan M. Vilar Fernández
A Doubly Smoothed PD Estimator in Credit Risk
Reprinted from: *Proceedings* **2020**, *54*, 55, doi:10.3390/proceedings2020054055 . . . . **174**

Alejandro Mosteiro Vázquez, Carlos Dafonte and Ángel Gómez
Open Source Monitoring System for IT Infrastructures Incorporating IoT-Based Sensors
Reprinted from: *Proceedings* **2020**, *54*, 56, doi:10.3390/proceedings2020054056 . . . . **178**

Macarena Díaz, Jorge Novo, Manuel G. Penedo and Marcos Ortega
Fully Automatic Method for the Visual Acuity Estimation Using OCT Angiographies
Reprinted from: *Proceedings* **2020**, *54*, 57, doi:10.3390/proceedings2020054057 . . . . **181**

Alejandro Lopez-Fernandez, Ruben Carneiro-Medin, Thais Pousada, Betania Groba-González and Adriana Dapena
Development of Recreational Content with Micro:Bit for Intervention with People with Cerebral Palsy
Reprinted from: *Proceedings* **2020**, *54*, 58, doi:10.3390/proceedings2020054058 . . . . **183**

Marta Moreno, Abel Sousa, Marta Melé, Rui Oliveira and Pedro G Ferreira
Predicting Gastric Cancer Molecular Subtypes from Gene Expression Data
Reprinted from: *Proceedings* **2020**, *54*, 59, doi:10.3390/proceedings2020054059 . . . . **187**

Carla Guerra Tort, Vanessa Aguiar Pulido, Victoria Suárez Ulloa, Francisco Docampo Boedo, José Manuel López Gestal and Javier Pereira Loureiro
Electronic Health Records Exploitation Using Artificial Intelligence Techniques
Reprinted from: *Proceedings* **2020**, *54*, 60, doi:10.3390/proceedings2020054060 . . . . **191**

Shuaiqiang Liu, Álvaro Leitao, Anastasia Borovykh and Cornelis W. Oosterlee
Machine Learning to Compute Implied Volatility from European/American Options Considering Dividend Yield
Reprinted from: *Proceedings* **2020**, *54*, 61, doi:10.3390/proceedings2020054061 . . . . **194**

About the Special Issue Editors

Joaquim de Moura, (Ph.D.) received his degree in Computer Engineering in 2014 from the University of A Coruña (Spain). In 2016, he received his M.Sc. degree in Computer Engineering from the same university and in 2019 his Ph.D. degree (Cum Laude) in Computer Science and Artificial Intelligence. He has also worked as a visiting researcher at the INESC-TEC (Portugal), in the development of different methodologies for DME diagnosis with OCT images. He is currently a postdoctoral researcher in the Center for Research in Information and Communication Technologies (CITIC) at the University of A Coruña. His main research interests include computer vision, biomedical image, and video processing.

Alejandro Puente-Castro (Ph.D. Student) gained his MSc. in Bioinformatics for Health Sciences and has worked in fields, such as early detection of Alzheimer's disease using Deep Learning techniques or selfquantification. Currently, his research is focused on applying Artificial Intelligence techniques to the coordination of heterogeneous groups of Unmanned Aerial Vehicles (UAVs).

Javier Pereira Loureiro (Ph.D.), associate professor at the Faculty of Health Science from the Universidade da Coruña (UDC), Spain. He received his B.Sc., Master degree and Ph.D. in Computer Science from the UDC in 1995, 1997 and 2004. Currently, he is deputy director of the Centre for Information and Communications Technology Research (CITIC) and the Head of Master of Bioinformatics. His research interests include medical informatics and assistive technologies for people with disabilities.

Manuel G. Penedo (Ph.D.) received the B.Sc. and Ph.D. degrees in Physics from the University of Santiago de Compostela, Spain, in 1990 and 1997, respectively. He is currently Professor at the Department of Computer Science in the University of A Coruña, Spain, and the director of the Center for Research in Information and Communication Technologies (CITIC). His main research interests include computer vision, biomedical image processing and video processing.

Preface to "3rd XoveTIC Conference"

A Coruña, Spain; 8–9 October 2020.

The 3rd XoveTIC Conference, organised by the University of A Coruña Centre for ICT Research (CITIC), is a platform for young pre- and post-doctoral researchers in the ICT area to present short papers and posters on the work they are doing. The aim of the conference is to complement and enhance young researchers' training by providing them with a space for scientific debate and exchange.

Joaquim de Moura , Alejandro Puente-Castro, Javier Pereira Loureiro , Manuel G. Penedo

Special Issue Editors

Proceedings

A First Approach to Authentication Based on Artificial Intelligence for Touch-Screen Devices †

Arturo Silvelo *, Daniel Garabato, Raúl Santoveña and Carlos Dafonte

CITIC, Department of Computer Science and Information Technologies, University of A Coruña, Campus de Elviña S/N, 15071 A Coruña, Spain; daniel.garabato@udc.es (D.G.); raul.santovena@udc.es (R.S.); dafonte@udc.es (C.D.)

* Correspondence: arturo.silvelo@udc.es

† Presented at the 3rd XoveTIC Conference, A Coruña, Spain, 8–9 October 2020.

Published: 18 August 2020

Abstract: Most authentication schemes follow a classical approach, where the users are authenticated only once at the beginning of their sessions. Therefore, it is not possible to verify the legitimate use of such a session or to detect any usurpation. In order to address this issue, we propose a second-phase authentication scheme that provides not only continuous user authentication during their sessions, but also in a transparent manner, since no additional or intrusive hardware is required. To this purpose, a novel approach was applied to create specific user profiles by means of different Artificial Intelligence techniques. In this work, we aim to study the feasibility of such an authentication scheme, so that it could be applied to a real time environment in order to verify the identity of the actual user against the legitimate user profile.

Keywords: authentication; artificial intelligence; cybersecurity

---

1. Introduction

Since the beginning of Information Technologies, authentication models have been an essential component for information security [1] and they have been adapted to the new devices and technologies that have appeared over the years, such as mobile phones, which are nowadays an indispensable tool to perform any daily operation. However, the authentication mechanisms commonly used present certain issues that can lead to security incidents related to weak or lost passwords. For these reasons, new and more secure authentication systems [2,3], such as the biometric ones, have been implemented, making use of unique human traits as passwords. Even so, these authentication systems continue to present a common problem, since they only verify the legitimacy of the user at the beginning of the session and not during it.

In this work, we propose a continuous authentication model which is based on the monitorization of the users' behavior [4] during the usage of a mobile device. Such a model is presented as a second authentication factor, which verifies the legitimacy of the user in a transparent manner, being able to detect if the user who is making use of the session is the one that was originally authenticated. For this purpose, it was necessary to create a multiplatform application that gathers data from the available motion sensors (mainly, accelerometer and gyroscope) and the touch-screen to generate a specific profile for each user by means of Artificial Intelligence (AI) techniques.

2. Methods

In order to conduct this experiment, we developed an application that collects information from the events associated with the use of the device. Those events related to motion sensors were grouped in time windows, whereas the touch-screen ones were grouped into gestures (swipe, rotate, tap, press, pinch, and pan) in order to seek for patterns to authenticate users. The data collected over three months were processed to extract a set of features, but first we needed to carefully analyze and treat such data in order to fit them to the classification techniques that were used. Some processes performed were the treatment of null and empty values, data standardization, definition of numerical/categorical variables, as well as feature extraction and selection.

Following this procedure, we obtained around 50 genuine features and many polynomial derived ones. Hence, feature selection was a key task in this phase and different techniques, such Random Forest or Recursive Feature Elimination processes, were used to assign different weights to the features, so that the most relevant ones could be identified in order to build the models.

Then we can feed some different well-known classification techniques (such as Random Forest [5], Support Vector Machines [6] and Multi-layer Perceptrons [7]) with the selected features in order to create a profile for each legitimate user in the system. These techniques require a training process so that they can appropriately fit the data, and the best configuration for each one was determined via hyper-parameterization and cross-validation procedures.

3. Results

Table 1 shows different common metrics used for classification tasks (accuracy, precision, recall and F1 Score). Among those metrics, we considered the precision as the most relevant one for our task, since it measures the number of intrusions into the system. These results show that the average response for our system is over 80% in terms of precision, and the F1 score is near 75%. The lowest score is obtained for the recall metric, which measures the identification of legitimate users as impostors.

Table 1. Results for each grouped by selected algorithms and events.

Users	Accuracy	Precision	Recall	F1 Score
01	80.29%	81.18%	78.85%	80.00%
02	78.11%	85.57%	68.78%	76.26%
03	80.17%	90.91%	69.67%	78.89%
04	81.82%	92.28%	70.92%	80.20%
05	79.62%	81.09%	76.90%	78.94%
06	72.97%	80.65%	64.10%	71.43%
07	52.50%	62.50%	43.48%	51.28%
08	63.12%	69.35%	51.81%	59.31%
09	68.94%	80.00%	59.46%	68.22%
10	84.07%	89.73%	78.30%	83.63%
11	83.75%	89.86%	76.54%	82.67%
12	71.92%	82.00%	56.16%	66.67%
13	76.19%	80.99%	68.45%	74.19%
14	82.85%	90.38%	75.20%	82.10%
15	81.43%	86.16%	76.11%	80.83%
16	76.07%	80.90%	69.76%	74.92%
17	77.84%	84.83%	68.72%	75.93%
18	76.85%	80.29%	69.62%	74.58%
Total	**76.03%**	**82.70%**	**67.94%**	**74.45%**

4. Conclusions and Future Work

As can be observed in Table 1, it has been demonstrated that a user's behavior can be used for authentication purposes. Although the recall metric shows that there is a considerable ratio of legitimate user misidentification, which may prevent this method as a primary authentication scheme, the obtained precision score encourages us as to its usage as a second authentication scheme that monitors user activity in a continuous and transparent manner over the entire session, since true impostors are usually detected. Such a system may have conservative behavior, and sometimes an alert would be raised for a legitimate user, but it only aims at detecting intrusions into the system once the user is already authenticated by a primary authentication scheme, such as a user/password one. Then, in case any usurpation is detected, the system could require the user to re-authenticate using the primary scheme, or even raise some notifications to the system administrators so that they can take further action.

In order to implement such an authentication scheme, we plan to use a streaming platform capable of handling events in near real time (sending, processing and storing the events for retraining against AI models (Figure 1)), so that the identity of the actual user can be verified in a short-time period.

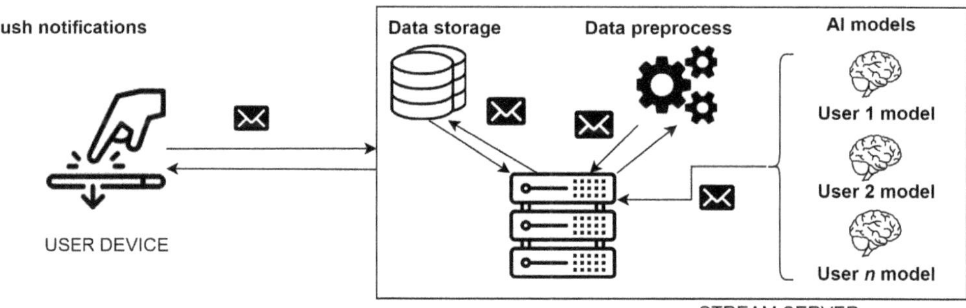

Figure 1. Authentication scheme for streaming platform.

Funding: This work has been mainly funded by the contracts CITIC-ACATIA, and CITIC-ACATIA2 established with the private entity ODEENE Ingeniería.

Acknowledgments: We also wish to acknowledge the support received from the CITIC as a Research Centre of the Galician University System is financed by the Conselleria de Education, Universidades e Formacin Profesinal (Xunta de Galicia) through the ERDF (80%), Operational Programme ERDF Galicia 2014–2020 and the remaining 20% by the Secretaria Xeral de Universidades (Ref. ED431G 2019/01), the research group fund ED431B 2018/42, and the Ministerio de Educación Cultura FPU16/03827 grant.

Conflicts of Interest: The authors declare no conflict of interest. The funders had no role in the design of the study; in the collection, analyses, or interpretation of data; in the writing of the manuscript, or in the decision to publish the results.

References

1. Stallings, W.; Brown, L. *Computer Security: Principles and Practice*, 1st ed.; Prentice Hall Press: Upper Saddle River, NJ, USA, 2007.
2. Barkadehi, M.H.; Nilashi, M.; Ibrahim, O.; Zakeri Fardi, A.; Samad, S. Authentication systems: A literature review and classification. *Telemat. Informatics* **2018**, *35*, 1491–1511, doi:10.1016/j.tele.2018.03.018.
3. Garabato, D.; García, J.; Nóvoa, F.; Dafonte, C. Mouse Behavior Analysis Based on Artificial Intelligence as a Second-Phase Authentication System. *Proceedings* **2019**, *21*, 29, doi:10.3390/proceedings2019021029.
4. ul Haq, M.E.; Awais Azam, M.; Naeem, U.; Amin, Y.; Loo, J. Continuous authentication of smartphone users based on activity pattern recognition using passive mobile sensing. *J. Netw. Comput. Appl.* **2018**, *109*, 24–35, doi:10.1016/j.jnca.2018.02.020.
5. Breiman, L. Random forests. *Mach. Learn.* **2001**, *45*, 5–32.
6. Cortes, C.; Vapnik, V. Support-vector networks. *Mach. Learn.* **1995**, *20*, 273–297.
7. Rumelhart; David, E.; James, M.; James, L. *Parallel Distributed Processing: Explorations in The Microstructure of Cognition. Volume 1. Foundations*; MIT Press: Cambridge, MA, USA, 1986.

Proceedings

Computation of Resonance Modes in Open Cavities with Perfectly Matched Layers †

Luis M. Hervella-Nieto , Andrés Prieto and Sara Recondo *

Research Group M2NICA, Department of Mathematics, CITIC, Universidade da Coruña, Campus de Elviña, 15071 A Coruña, Spain; luis.hervella@udc.es (L.M.H.-N.); andres.prieto@udc.es (A.P.)

* Correspondence: sara.recondo@udc.es

† Presented at the 3rd XoveTIC Conference, A Coruña, Spain, 8–9 October 2020.

Published: 18 August 2020

Abstract: During the last decade, several authors have addressed that the Perfectly Matched Layers (PML) technique can be used not only for the computation of the near-field in time-dependent and time-harmonic scattering problems, but also to compute numerically the resonances in open cavities. Despite such complex resonances are not natural eigen-frequencies of the physical system, the numerical determination of this kind of eigenvalues provides information about the model, what can be used in further applications. The present work will be focused on two main specific goals—firstly, the mathematical analysis of the frequency-dependent highly non-linear eigenvalue problem associated to the computation of resonances with the standard PML technique. Second, the implementation of a robust numerical method to approximate resonances in open cavities.

Keywords: exterior acoustics; perfectly matched layers (PML); open cavities

1. Introduction

The calculation of resonances has been always relevant in applied acoustics as well as in the research field. Its application is not only very extensive but critical, for instance, in the study of many mechanical systems. Moreover, the solution to this kind of problems leads to some singularities that keep different research lines open in this field. More specifically, the eigenvalue problem in exterior acoustics has been deeply studied during the last years, due to its difficulty compared to its application over bounded domains (see for instance [1,2]). For instance, the computation of leaky modes in an open waveguide can be used as the spectral basis for a modal decomposition technique [3].

There are two main approaches to this type of problem. The first one includes the boundary element method (BEM), based on the solution by surface integration over an interior boundary. This technique has certain computational advantages due to the dimensional reduction of the computational domain, but requires very sophisticated numerical integration techniques, as analyzed in Reference [4]. Another known group of methods that has become relevant in the last years is based on the discretization of the infinite domain. Examples of this type of approach are the combination of Finite Elements Method (FEM) with Infinite Elements Method (IFEM) [5,6], as well as with Absorbing Boundary Conditions (ABCs) [7] or Perfectly Matched Layers (PML) [8–10].

All these techniques are based on the division of the problem into two different domains, thus being able to work with an interior and an exterior one. The exterior one simulates the conditions of an infinite domain around the region of interest, contained in the interior one. The main difference between these types of methods lies in the way that this second domain is handled. This work will be focused on the use of PMLs, which consists in surrounding the interior region of interest with some sponge layers which couple perfectly without generating any spurious reflections. The key ingredient of the PML technique is the definition of an adequate absorption profile which allows to damp the outgoing waves travelling at free field regime from the interior region of interest [8]. So, this type of exterior PML domain is characterized by its thickness and the associated absorption profile. In [10] a variety of scenarios are analyzed, showing that the numerical performance of the PML technique depends clearly on the data set, the mesh and the geometry of the problem.

2. Expected Results

The proposed work will be devoted to two main goals. First, the mathematical analysis of the PML technique involving non-integrable absorption profiles will be faced, to analyze whether the computed eigenvalues are independent of the thickness of the PML, as shown in Reference [9] for source problems. Secondly, a robust numerical method will be implemented in order to avoid the instabilities presented by the use of finite element methods, as shown in Reference [1], for resonance calculation in open cavities. The proposed method will try to avoid dependencies with respect to the PML thickness, as well as the position of the inner PML boundary, which is shared with the physical interior region of interest. More precisely, a finite element discretization (or other high order methods) will be implemented to illustrate the numerical behavior associated with different discretizations used in the computation of resonances. The observed numerical instabilities in the different discrete methods will be analyzed, taking into account that the linear matrices involved in the discretization are not hermitian. Once the origin of such instabilities are identified, the current state-of-the-art non-linear eigenvalue techniques will be used to design a robust methodology.

Funding: This work has been supported by MTM project PID2019-108584RB-I00 (MICINN), accreditations ED431C 2018/33 - M2NICA (Xunta de Galicia & ERDF) and ED431G 2019/01 - CITIC (Xunta de Galicia & ERDF).

References

1. Hein, S.; Hohage, T.; Koch, W.; Schöberl, J. Acoustic resonances in a high-lift configuration. *J. Fluid Mech.* **2007**, *582*, 179.
2. Aguilar, J.; Combes, J.M. A class of analytic perturbations for one-body Schrödinger Hamiltonians. *Commun. Math. Phys.* **1971**, *22*, 269–279.
3. Dhia, A.B.B.; Goursaud, B.; Hazard, C.; Prieto, A. Finite element computation of leaky modes in stratified waveguides. In *Ultrasonic Wave Propagation in Non Homogeneous Media*; Springer: Berlin/Heidelberg, Germany, 2009; Volume 128, pp. 73–86.
4. Kress, R. Boundary integral equations in time-harmonic acoustic scattering. *Math. Comput. Model.* **1991**, *15*, 229–243.
5. Astley, R. Infinite elements for wave problems: a review of current formulations and an assessment of accuracy. *Int. J. Numer. Methods Eng.* **2000**, *49*, 951–976.
6. Moheit, L.; Marburg, S. Convergence of modes in exterior acoustics problems with infinite elements. In Proceedings of the 45th International Congress and Expositionon Noise Control Engineering, Hamburg, Germany, 21–24 August 2016.
7. Turkel, E. Introduction to the special issue on absorbing boundary conditions. *Appl. Numer. Math.* **1998**, *4*, 327–329.
8. Berenger, J.P. A perfectly matched layer for the absorption of electromagnetic waves. *J. Comput. Phys.* **1994**, *114*, 185–200.
9. Bermúdez, A.; Hervella-Nieto, L.; Prieto, A.; Rodrı, R. An optimal perfectly matched layer with unbounded absorbing function for time-harmonic acoustic scattering problems. *J. Comput. Phys.* **2007**, *223*, 469–488.
10. Bermúdez, A.; Hervella-Nieto, L.; Prieto, A.; Rodríguez, R. An exact bounded perfectly matched layer for time-harmonic scattering problems. *SIAM J. Sci. Comput.* **2008**, *30*, 312–338.

Proceedings

SARDAM: Service Assistant Robot for Daily Activity Monitoring †

Celtia Meizoso Lamas ‡, Francisco Bellas *,‡ and Berta Guijarro-Berdiñas ‡

Centre for Information and Communications Technology Research (CITIC), University of A Coruña, Campus de Elviña, s/n, 15071 A Coruña, Spain; c.mlamas@udc.es (C.M.L.); berta.guijarro@udc.es (B.G.-B.)

* Correspondence: francisco.bellas@udc.es

† Presented at the 3rd XoveTIC Conference, A Coruña, Spain, 8–9 October 2020.

‡ These authors contributed equally to this work.

Published: 18 August 2020

Abstract: In this work, we propose an autonomous monitoring system for the daily routine of an elderly person. SARDAM (Service Assistant Robot for Daily Activity Monitoring), which is the name of this system, uses a humanoid robot as a key element that carries out a direct interaction with the user. The purpose of SARDAM is to keep the user active as long as possible by suggesting, and monitoring, a series of daily tasks and healthy habits according to the prescription of a health professional, in order to reduce the early appearance of cognitive and motor impairment. In the current version of SARDAM we use the NAO humanoid robot, which performs a natural interaction with the user through vision and speech libraries. To assure the appropriate execution of the user's daily tasks, a module for emotion detection has been incorporated in order to propose corrective tasks according to the detected emotion. SARDAM was tested in a scenario with a real user, getting successful results and positive opinions from them that encourage further work.

Keywords: socially assistive robotics; artificial intelligence; NAO robot; human-robot interaction; emotion detection; object detection; speech recognition

1. Introduction

Socially Assistive Robotics (SAR) is a currently booming field due to the need for taking care of the growing number of elderly people and people who needs special treatment. One of the main goals of SAR is to maintain an independent lifestyle in the users as long as possible with the help of a robot. These type of robots provide physical therapy and daily assistance, as well as emotional support to the user, among other functions [1]. As for the use of SAR in elderly care, physical and cognitive care, monitoring and training is provided by the robots to improve the quality of life of the people [2].

Presently, there are some relevant projects in this field, such as Pearl [3], from Nursebot project, which mainly reminds the user of the daily activities to be carried out, and guides him/her through the environment. Another SAR system is ENRICHME [4], which uses an customized version of the TIAGo robot, and its main goal is to provide home assistance, health monitoring, complementary care, and social support to improve the quality of life of the elderly person.

SARDAM, the prototype proposed in this paper, is a monitoring system of the daily routine of an elderly person that uses the NAO robot. It is based on a set of activities that are executed following a daily schedule, which is customized for the user, and which is organized depending on a set of priorities as established by a health professional. SARDAM's main goal is to remind the user of the activities he/she has to do at the right time, emphasising those with the highest priority, and using other activities as a resource to maintain an optimal routine. SARDAM suggests the activities according to the user's emotional state, adding, when necessary, some special activities to improve it and encourage the user to follow the proper routine.

2. SARDAM's Architecture

Figure 1 shows SARDAM's architecture, which consists of five main modules. Starting from the top, the user and NAO robot interact with each other. From this interaction, the robot's sensors data are passed to the detectors, which process the information. These detectors are focused on people detection and recognition, speech recognition, face emotion recognition, touch detection, and object detection. After that, the processed data are used by the activity monitoring system and the emotion recognition system. The activity monitoring system executes the daily schedule and registers data about the person's activity on a log file. During the execution of the activities, it provides orders to the robot's actuators on what they have to do depending on the interaction required. Such actuators are in charge of NAO's behaviour. The suggestion of each activity depends on the person's emotional state. If it is detected that the user is not feeling well enough to execute the activity, then the system will try to encourage him/her with a corrective activity during a short time period.

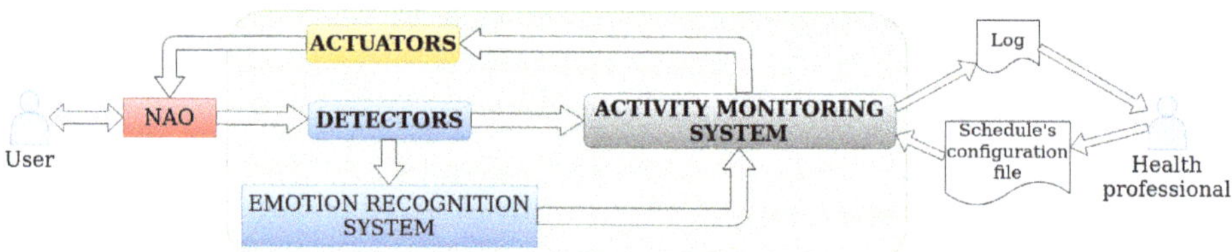

Figure 1. SARDAM's architecture.

2.1. The Emotion Recognition System

The emotion recognition system predicts the user's mood and provides it to the activity monitoring system every time it is needed. To obtain such mood prediction, it merges the results of emotion recognition on frames of a video and on the detected speech from an audio. The library used to perform face emotion recognition is FaceEmotion_ID [5], while SpeechRecognition [6] is the Python library used to detect the speech from an audio, using Google Web Speech API.

To obtain the results from video, 10 frames are extracted to label emotions on each of them using FaceEmotion_ID. Once such labels ("Good mood", "Normal" and "Bad mood") with their confidence are obtained, they are merged to get the final results of face emotion. As for results from the audio, first a speech recognition is performed. The next step is to label each detected word using the already mentioned three labels. To do this, it was established a bag of words of each label. Their confidence values are also calculated.

Finally, the definitive label that qualifies the user's emotional state is calculated from the average of the results obtained from the video and the audio.

2.2. The Activity Monitoring System

The activity monitoring system is in charge of loading and executing the schedule which contains the activities of the user's daily routine, wrote by a health professional on a configuration file. Currently, it is composed by 18 activities of a typical daily routine for an elderly person, organized on three levels of priority. For instance, task of the highest priority are "wash up", "eating" or "physical activity", while task of intermediate priority are "clean the house" or "calling a friend".

3. Experiment on a Real Environment

SARDAM was tested in a real environment through a one-day monitoring experiment in a home with a woman in her 60s. First of all, it was required to set the daily activities in the schedule configuration file. Once the system started, the schedule was executed as follows. First, SARDAM warns the user what is the task to be carried out; for example, "you should go to have breakfast". If the task is performed, nothing happens until the next one in the schedule. However, if the user does not perform it, SARDAM uses the emotion recognition system to decide which corrective activity is proposed. Once this special activity is finished, the previously postponed task is proposed again.

As this experiment was performed during a whole day, it was possible to test how the system works dealing with different user responses. We overall got successful results, although in some cases, the detected mood was not the real one expressed by the user. At the end of the day, we asked the user about her experience with the robot, and she expressed that she felt that the robot really cared for her, and that it has been really useful having such a partner reminding her to follow her daily routine.

Figure 2 shows two images retrieved during the execution of this experiment. For more examples about SARDAM's performance: https://www.youtube.com/watch?v=Rz10ANdjpy8.

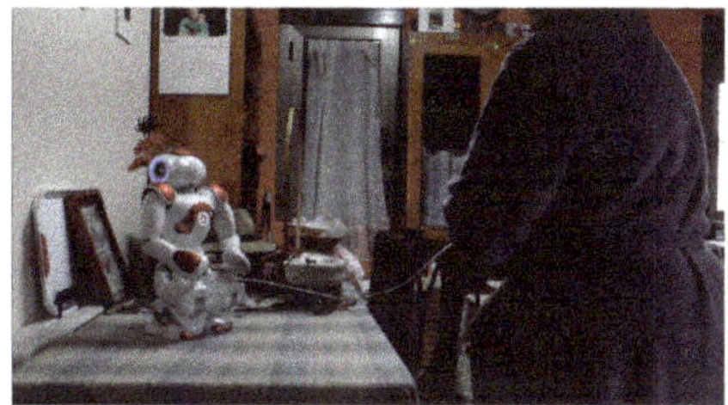

(**a**) Talking to the user in the wake up activity.

(**b**) Execution of the physical activity.

Figure 2. Images retrieved from the execution of the experiment with a real user.

4. Conclusions and Future Work

SARDAM is a Socially Assistive Robotics system which is in charge of monitoring the daily routine of an elderly person using emotion recognition. Preliminary results have been encouraging, although many improvements can be performed in the future, like improving the emotion recognition system and incorporating a schedule's replanification module which adapts to the user's preferences.

Conflicts of Interest: The authors declare no conflict of interest.

References

1. Feil-Seifer, D.; Matarić, M. Defining Socially Assistive Robotics. In Proceedings of the 9th International Conference on Rehabilitation Robotics, 2005 (ICORR 2005), Chicago, IL, USA, 28 June–1 July 2005; Volume 2005, pp. 465–468.
2. Broekens, J.; Heerink, M.; Rosendal, H. Assistive social robots in elderly care: A review. *Gerontechnology* **2020**, *8*, 94–103.
3. Pollack, M.; Brown, L.; Colbry, D.; Orosz, C.; Peintner, B.; Ramakrishnan, S.; Engberg, S.; Matthews, J.; Dunbar-Jacob, J.; Mccarthy, C.; et al. Pearl: A Mobile Robotic Assistant for the Elderly. In Proceedings of the AAAI Workshop on Automation as Eldercare, Edmonton, AB, Canada, 28 July 2002.
4. Coşar, S.; Fernández-Carmona, M.; Agrigoroaie, R.; Pages, J.; Ferland, F.; Zhao, F.; Yue, S.; Bellotto, N.; Tapus, A. ENRICHME: Perception and Interaction of an Assistive Robot for the Elderly at Home. *Int. J. Soc. Robot.* **2020**, *12*, 779–805.
5. FaceEmotion_ID. Available online: https://github.com/abhijeet3922/FaceEmotion_ID (accessed on 23 July 2020).
6. SpeechRecognition. Available online: https://pypi.org/project/SpeechRecognition/ (accessed on 23 July 2020).

Proceedings

A Collaborative Augmented Reality Application for Training and Assistance during Shipbuilding Assembly Processes †

Aida Vidal-Balea [1,2,*], Oscar Blanco-Novoa [1,2], Paula Fraga-Lamas [1,2,*], Miguel Vilar-Montesinos [3] and Tiago M. Fernández-Caramés [1,2,*]

1 Department of Computer Engineering, Faculty of Computer Science, Universidade da Coruña, 15071 A Coruña, Spain; o.blanco@udc.es
2 Centro de Investigación CITIC, Universidade da Coruña, 15071 A Coruña, Spain
3 Navantia S. A., Astillero de Ferrol, 15403 Ferrol, Spain; mvilar@navantia.es
* Correspondence: aida.vidal@udc.es (A.V.-B.); paula.fraga@udc.es (P.F.-L.); tiago.fernandez@udc.es (T.M.F.-C.); Tel.: +34-981167000 (ext. 6051) (P.F.-L.)

† Presented at the 3rd XoveTIC Conference, A Coruña, Spain, 8–9 October 2020.

Published: 18 August 2020

Abstract: This paper presents the development of a novel Microsoft HoloLens collaborative application that allows shipyard operators to interact with a virtual clutch during its assembly in a real Turbine workshop. Such an Augmented Reality (AR) experience acts as a virtual guide while assembling different parts of a ship. In particular, the proposed application allows operators to position the clutch on a real environment and interact with it. The application also provides information about the documentation of each part of the clutch, showing its blueprints and physical measurements. The proposed AR application enables collaborative AR experiences, allowing users to visualize the same content and animations at the same time and interact simultaneously with 3D objects from multiple devices. Furthermore, the application is integrated with an Industrial Internet of Things (IIoT) framework, resulting on an AR-IIoT application that is able to receive and display real-time sensor data on information panels, as well as to trigger actions through actuators by making use of virtual user interfaces.

Keywords: Industry 4.0; augmented reality; industrial augmented reality; internet of things; training; collaborative application

---

1. Introduction

Shipbuilding companies are updating and integrating new technologies on their working processes leading to what is called Shipyards 4.0. Among the different Industry 4.0 technologies that can be used in a Shipyard 4.0, one of the most promising is Augmented Reality (AR), which provides a wide range of features that can be leveraged in order to develop powerful and useful applications that project virtual elements into real scenarios. These features enable improving industrial processes, maximizing worker efficiency, as well as incorporating new learning and training methods that can be integrated on a shipbuilding company.

The objective of this work is to study the potential of Industrial Augmented Reality (IAR) to facilitate, support and optimize production and assembly tasks. With this aim, an IAR application has been developed with Microsoft HoloLens glasses [1] to guide operators and to facilitate training tasks in a visual and practical way.

2. Materials and Methods

2.1. Hardware and Software

The proposed IAR application was developed using two Microsoft HoloLens 1st generation smart glasses. The application development was carried out with Unity 2019.3.3f1 using HoloToolkit 2017.4.3.0, as well as Visual Studio 2017 for building and deploying the Unity project into the HoloLens glasses. In addition, in order to be able to compile the project for Universal Windows Platform (UWP), Windows 10 Creators Update and Microsoft UWP Software Development Kit (SDK) were used. All the 3D models were processed with Blender 2.8.

2.2. Methodology

For the development of the IAR application an iterative methodology was followed, in which the different functionalities were added incrementally. In the first iteration the basic functionality was implemented, including the visualization of gear movements and assembly animations, the addition of simple buttons to perform actions, User Interface (UI) translation and 3D model simplification (polygon reduction). In the second iteration the User Experience (UX) was improved, replacing the buttons with a panel that makes use of the holographic buttons provided by HoloToolkit and adding billboard and tag-along functionality to the panel (this implies that the panel follows the user and it is always facing towards his/her direction). In the third iteration the documents associated with the production order and the component blueprints were added, so that when the user taps on them, a panel is shown with such a documentation. In the fourth iteration the shared experience was implemented and integrated. Finally, in the fifth iteration, an Industrial Internet of Things (IIoT) framework was integrated [2], including the corresponding panel modifications, from which the connected sensor values can be seen and interacted with.

3. Results

Figure 1 shows, on the left, a screenshot of the developed IAR application where a step sequence for the correct assembly of a clutch is visualized through animations and together with the blueprints and documentation related to its components. The application can be executed simultaneously and in a synchronized way by different smart glasses that are in the same physical location, so all users see the same virtual objects and animations at the same time and in the same position. In addition, the application is capable of receiving and displaying information from sensors in real time as well as triggering actions through actuators by making use of a virtual user interface.

Figure 1. Interaction with the 3D model (**left**) and assembly process at Navantia's Turbine workshop (**right**).

The developed solution was tested in the Turbine workshop of Navantia in Ferrol during the assembly process of a clutch (illustrated in Figure 1 on the right), which consisted of a high number of mobile parts with varied shapes and sizes. Its aim was to optimize the process by reducing assembly time and the risks associated with human errors. Therefore, the proposed IAR application allows for illustrating visually the contents of the assembly manuals that are currently printed on paper, connecting the clutch with external IIoT devices and as well as easing collaborative training tasks.

4. Discussion

The proposed IAR application was designed and tested for training and assistance during shipbuilding assembly. The developed functionalities allow for visualizing virtual elements in detail at their real size, placed in a precise way in a position where the real elements would be integrated, even before the production of such elements starts. Moreover, the proposed application enables IIoT interactions, collaborative training and assistance tasks to speed up manufacturing processes.

Funding: This work was supported by the Plant Information and Augmented Reality research line of the Navantia-UDC Joint Research Unit and has been funded in part by Xunta de Galicia (IN853B-2018/02).

Conflicts of Interest: The authors declare no conflicts of interest.

References

1. HoloLens Official Web Page. Available online: https://www.microsoft.com/en-us/hololens (accessed on 13 July 2020).
2. Blanco-Novoa, Ó.; Fraga-Lamas, P.; A. Vilar-Montesinos, M.; Fernández-Caramés, T.M. Creating the Internet of Augmented Things: An Open-Source Framework to Make IoT Devices and Augmented and Mixed Reality Systems Talk to Each Other. *Sensors* **2020**, *20*, 3328.

Proceedings

Implementing a Web Application for W3C WebAuthn Protocol Testing †

Martiño Rivera Dourado [1,*], Marcos Gestal [1,2] and José M. Vázquez-Naya [1,2]

1 Grupo RNASA-IMEDIR, Departamento de Computación, Facultade de Informática, Universidade da Coruña, Elviña, 15071 A Coruña, Spain; marcos.gestal@udc.es (M.G.); jose.manuel.vazquez.naya@udc.es (J.M.V.-N.)
2 Centro de Investigación CITIC, Universidade da Coruña, Elviña, 15071 A Coruña, Spain
* Correspondence: martino.rivera.dourado@udc.es

† Presented at the 3rd XoveTIC Conference, A Coruña, Spain, 8–9 October 2020.

Published: 18 August 2020

Abstract: During the last few years, the FIDO Alliance and the W3C have been working on a new standard called WebAuthn that aims to substitute the obsolete password as an authentication method by using physical security keys instead. Due to its recent design, the standard is still changing and so are the needs for protocol testing. This research has driven the development of a web application that supports the standard and gives extensive information to the user. This tool can be used by WebAuthn developers and researchers, helping them to debug concrete use cases with no need for an *ad hoc* implementation.

Keywords: WebAuthn; authentication; testing

1. Introduction

Authentication is one of the most critical parts of an application. It is a security service that aims to guarantee the authenticity of an identity. This can be done by using several security mechanisms but currently, without a doubt, the most common is the username and password method. Although this method is easy for a user to conceptually understand, it constitutes many security problems. Some of them are password stealing through phishing and leaked password hash cracking by using dictionary attacks [1]. User passwords can also be compromised at the client side by the use of malware or hardware devices. For this purpose, there exist some techniques that capture the password on user input, often involving keyloggers [2]. All these attacks leave the user unprotected and without control over their credentials.

One of the approaches that big technology companies such as Facebook or Google started to implement is based on the use of hardware authenticators [3], such as the Yubikey [4] or the Titan Security Key [5]. This idea evolved into a web authentication protocol, called WebAuthn, which was published in March 2019 as a W3C Recommendation [6] as a result of some collaborations with the FIDO Alliance. Aside from the authenticator, the Relying Party, an "entity whose web application utilizes the Web Authentication API to register and authenticate users [6], must support the validation of the authenticator device responses.

Nowadays, although the standard has become a W3C Recommendation last year, the consortium is already building a second version, currently as a Working Draft, that corrects, improves and adds functionalities with respect to its first version. Therefore, there is an important need for testing the protocol. In many cases, a test environment can be useful, avoiding the need of an *ad hoc* implementation for reproducing a use case. For this reason, this research drove to the implementation of a web tool that allows developers and testers to debug browsers and authenticators compatible with the standard.

2. Tool Development

The developed tool displays the performed operations together with all necessary configuration details, as well as showing the researcher all the information exchanged through the web browser between the authenticator and the Relying Party. In order to provide a flexible test environment, the tool does not require a user account for its use. Built as Single Page Application and by using asynchronous requests to a web server, it allows to temporarily register credentials that can afterwards be authenticated for testing purposes. The server holds the minimum information required for the operations validation and can be accessed and deleted at the user interface. Moreover, the checks for browser compatibility and all errors occurred at the browser during any operation will be displayed at the interface. This approach gives the researcher both the server and the client data during the whole operation in a single web application.

The tool offers two main operations that follow the schema described in the Figure 1: registration and authentication. Once the user selects the operation, the web application will request the options for the selected WebAuthn operation, called Attestation for registration and Assertion for authentication (steps 0–1). In both cases, they include a challenge, which is a binary buffer generated and stored at the server. Once these default configurations are received, they can be edited by the user through a reactive form, allowing the researcher to configure settings such as requesting registration with resident credentials or changing the cryptographic algorithms. After this edition, the tool will interface with the authenticator through the WebAuthn API implemented in the compatible browser, sending these configurations to obtain the authenticator response after the user's physical interaction (step 2).

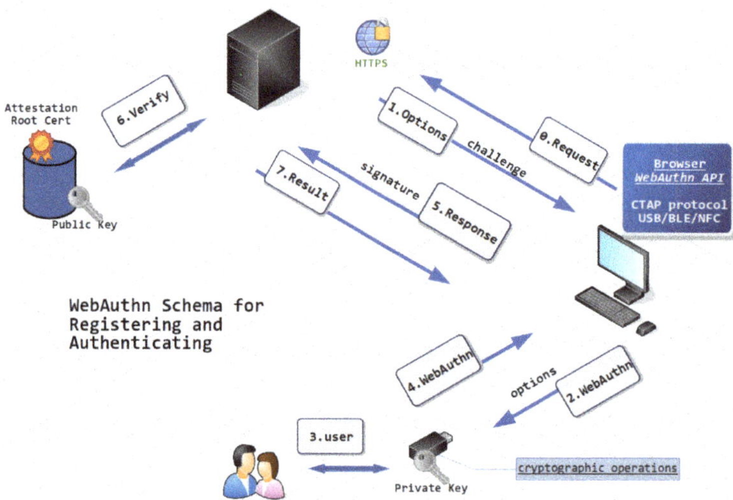

Figure 1. WebAuthn schema used by the developed tool for registration and authentication operations.

The authenticator response (step 4) is represented in some binary buffers. Like the challenge, these are encoded in Base64Url and sent to the server for performing validation (steps 5–6). During registration, once the Attestation is validated, the public key included in the authenticator response is stored at the server. On the other hand, during authentication, the public key of a registered credential will be used to validate the signature. In both operations, the result is forwarded to the web application together with all technical details, including the parsed authenticator response (step 7).

3. Results

DebAuthn (available at debauthn.tic.udc.es) is a WebAuthn debugging tool. It is a Single Page Application built with VueJS that stores temporary information tight to the session. The dashboard of the application displays the registered credentials—their counters, public keys and ids. Also, the user can delete all credentials in order to clean the workspace.

By using REST endpoints with NodeJS, the web application can send and request the necessary information, while displaying it to the user. The displayed data is divided in three steps corresponding to the main schema of the protocol: sending the options, authenticator response and server validation.

4. Conclusions

The developed tool can help both protocol researchers and developers. During the development of the new WebAuhn version, there are many contributions where a testing tool may be of use to reproduce a specific protocol use case. Furthermore, developers designing an implementation of the protocol in their systems may find the application as a useful tool for compatibility checks on authenticators and browsers, guiding their implementation decisions beforehand.

Author Contributions: Conceptualization, M.R.D.; Methodology, M.R.D., M.G. and J.M.V.-N.; Software, M.R.D.; Investigation, M.R.D., M.G. and J.M.V.-N.; Resources, M.G. and J.M.V.-N.; Writing—original draft preparation, M.R.D.; Writing—review and editing, M.R.D., M.G. and J.M.V.-N.; Supervision, M.G. and J.M.V.-N. All authors have read and agreed to the published version of the manuscript.

Funding: This work was supported by the Consolidation and Structuring of Competitive Research Units—Competitive Reference Groups (ED431C 2018/49), funded by the Ministry of Education, University and Vocational Training of the Xunta de Galicia endowed with EU FEDER funds.

Acknowledgments: We thank CITIC Research Center for hosting the developed tool.

Conflicts of Interest: The authors declare no conflict of interest.

Abbreviations

The following abbreviations are used in this manuscript:

FIDO	Fast Identity Online
W3C	World Wide Web Consortium
API	Application Programming Interface

References

1. ScatteredSecrets.com. How to Crack Billions of Passwords? Available online: https://medium.com/@ScatteredSecrets/how-to-crack-billions-of-passwords-6773af298172 (accessed on 18 April 2020).
2. Ahmed, Y.A.; Maarof, M.A.; Hassan, F.M.; Abshir, M.M. Survey of Keylogger Technologies. *IJCST* **2014**, *5*, 25–31.
3. Krebs, B. Google: Security Keys Neutralized Employee Phishing. Available online: https://krebsonsecurity.com/2018/07/google-security-keys-neutralized-employee-phishing/ (accessed on 18 April 2020).
4. Discover YubiKeys—Strong Two-Factor Authentication for Secure Login. Available online: https://www.yubico.com/products/ (accessed on 18 April 2020).
5. Titan Security Key Fido U2 F Usb C Nfc Ble. Available online: https://cloud.google.com/titan-security-key/ (accessed on 18 April 2020).
6. W3C. Web Authentication: An API for Accessing Public Key Credentials Level 1. Available online: https://www.w3.org/TR/webauthn/ (accessed on 18 April 2020).

Proceedings

Use of Arduino Microcontroller in Education: Creation of "The Musical Stairs" †

Felpeto-Guerrero Abraham, Vázquez-Sánchez Rubén, Chao-Fernández Rocío * and Chao-Fernández Aurelio

IMETIC Methodological Innovation through TIC, Faculty of Education, University of A Coruña, 15071 Elviña, Spain; a.felpeto@udc.es (F.-G.A.); ruben.vazquez.sanchez@udc.es (V.-S.R.); aurelio.chao@udc.es (C.-F.A.)

* Correspondence: rocio.chao@udc.es

† Presented at the 3rd XoveTIC Conference, A Coruña, Spain, 8–9 October 2020.

Published: 18 August 2020

Abstract: In the last few years, the introduction and use of Information and Communication Technology (onwards ICT) in the classroom has been gradually increased, especially in Science, Technology, Engineering, and Mathematics (onwards STEM) areas. However, the use of ICT technology within the music classroom seems to have stalled in most cases, being relegated to making sound recordings and listening to music fragments. Thanks to the rise of the Maker movement, more and more teachers are interested in introducing new types of ICT in the classroom. In our case, we will show how we developed "The Musical Stairs" to teach the musical concept of pitch, through the creation of a project based in Arduino.

Keywords: ICT in music classroom; educational technology; educational innovation; Arduino microcontroller; maker movement

---

1. Introduction

The role of teachers is increasingly bound to the use and integration of ICT within the teaching–learning process. Originally, the use of ICT in the classroom was mainly focused towards the introduction of computers, digital boards and the internet, where depending on their level of integration, the displacement of traditional or analogue audio–visual materials towards digital ones was, to some extent, noticeable [1]. However, thanks to the rise of the Maker movement, we are witnessing the inclusion of new types of technology and tools addressed to the classroom, as well as the proliferation of a new teacher profile, usually bound to STEM subjects, which rely on making as a way to reflect on the educational practice through research [2].

With regard to the musical education field, several studies show that a high percentage of teachers, although they think that ICT strengthens the teaching–learning process, still lack initial and permanent training regarding their use, therefore, although they have enough ICT resources in the classroom, they lack the necessary skills to use and to adapt them to their teaching methodology [3].

In this paper, we will show the creation process of "The Musical Stairs", under the assumption that, despite a teacher´s low ICT literacy rate, and thanks to the proliferation of the Maker movement as well as all its related features—user communities, development platforms, low cost hardware components, etc.—the music teacher will, up to a reasonable point, be able to be immersed within the Educative Robotics (ER) environment, allowing students to acquire musical competences and skills—in relation to other curricula subjects—from the perspective that the ER environment provides support and learning means [4], without underestimating the ludic, collaborative and motivational aspects inherent to the ER methodology.

2. Mounting Process and Materials

The "Musical Stairs" had to show the musical concept of pitch, therefore, through ascending or descending steps will show a simile with the ascending or descending pitch of a musical scale. In this way, the already said concept could be easily interiorized, among others [5].

The "Musical Stairs" principle of operation has been based on the reflection produced when a sonic burst triggered from an ultrasonic sensor collides with an object and returns to the sensor whenever the object is within the sensor range. In this case, it used the well-known 4-pin HC-SR04 sensor—power (VCC), ground (GND), trigger (Trig) and receptor (Echo)—.

To assemble the "Musical Stairs", first we covered each step with black and white vinyl as if it were a piano keyboard. Next, the sensors were placed along the stairs, disposing 12 or 7 sensors depending on if the preference was to represent a natural or chromatic scale, serially connected for both VCC and GND pins. The remaining pins were connected individually to an Arduino Mega microcontroller, to be able to control independently each sensor's input (Trig) and output (Echo). After that, the Arduino Mega microcontroller was programmed in a way that every 60 milliseconds a sonic burst is triggered and reflected. Once the data from the elapsed time between the sent and the received pulse—expressed in milliseconds—are received, it is time to calculate the distance of the object on which the burst collides, being in this case either a foot or a wall. For this, the formula S = d/t, where the resulting formula to know the distance in centimetres would be d(cm) = t(μs)/58. After that, the value resulting from the calculation of each sensor is serially sent to a laptop.

On the laptop, a development environment called Processing 3.0 was installed, along with the Minim library—an application programming interface (API) of a Java Sound library that allows for the integration of audio functions to the project. In our case, Processing is used to receive the measures from each of the sensors and associate each of them with a predetermined sound, in this case a musical note pre-recorded expressly for the project. In the giving case of receiving a measurement from a sensor less than 300 cm—which is the distance between each sensor and the wall—Processing plays their related sound until the measurement returns to 300 cm. The latter logic works for a musical natural scale of 7 sounds. If the purpose is to use a chromatic scale of 12 sounds—simulating the black and white keys of the piano—as the black keys are shorter than the white ones, their sound should be played when the detection distance is lesser than 100 cm. For this reason, within the source code it is necessary to differentiate between the black and white keys to be able to evaluate the distances and play the sounds depending on the position of each sensor. Figure 1 shows the operating diagram of all the above descripted.

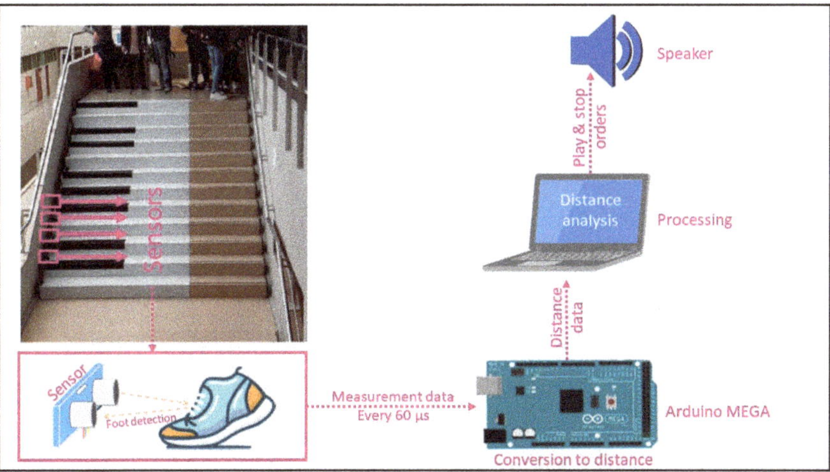

Figure 1. Operating diagram of "The Music Stairs" (source: the authors).

3. Conclusions

After carrying out this experience, we observed a change of mind in the future teachers—the project was developed in an education faculty—so that they are now more inclined and open to perform innovative projects using ICT, in addition to confirming a greater interest in the knowledge and development of their digital skills.

Author Contributions: Organization, C.-F.R. and C.-F.A.; implementation, C.-F.R., C.-F.A., F.-G.A., V.-S.R.; programming, F.-G.A.; writing, F.-G.A.; review and editing, C.-F.R., F.-G.A., V.-S.R. and C.-F.A. All authors have read and agreed to the published version of the manuscript.

Funding: This project was funded by the Faculty of Education from the University of A Coruña through the competitive call of grants for cultural activities.

Conflicts of Interest: The authors declare no conflict of interest.

References

1. Area-Moreira, M.; Hernández-Rivero, V.; Sosa-Alonso, J. Modelos de integración didáctica de las TIC en el aula. *Comunicar* **2016**, *24*, 79–87.
2. Vossoughi, S.; Bevan, B. Making and Tinkering: A Review of the Literature. *Natl. Res. Counc. Comm. Out Sch. Time Stem* **2014**, *2014*, 1–55.
3. Colás Bravo, M.P.; Hernández Portero, G. Itinerarios formativos del profesorado de música: sus percepciones sobre el valor didáctico de las TIC. *Fuentes* **2017**, *19*, 39–56.
4. Moreno, I.; Muñoz, L.; Rolando, J.; Quintero, J.; Pitti, K.; Quiel, J. La robótica educativa, una herramienta de enseñanza-aprendizaje de las ciencias y las tecnologías. *Tesi* **2012**, *13*, 74–90.
5. Chao-Fernández, R.; Felpeto-Guerrero, A.; Vázquez-Sánchez, R. Experiencia de creación y puesta en funcionamiento de la "escalera musical" como medio saludable de relacionar áreas de conocimiento. *IV Xornadas de Innovación Docente* **2020**, in press.

Proceedings

Regression Tree Based Explanation for Anomaly Detection Algorithm †

Iñigo López-Riobóo Botana [1,*], Carlos Eiras-Franco [2] and Amparo Alonso-Betanzos [2]

1 Research group LIDIA, Universidade da Coruña, Campus Elviña, 15071 A Coruña, Spain
2 CITIC Research Center, Universidade da Coruña, 15071 A Coruña, Spain; carlos.eiras.franco@udc.es (C.E.-F.); ciamparo@udc.es (A.A.-B.)
* Correspondence: inigo.lopezrioboo.botana@udc.es

† Presented at the 3rd XoveTIC Conference, A Coruña, Spain, 8–9 October 2020.

Published: 18 August 2020

Abstract: This work presents EADMNC (Explainable Anomaly Detection on Mixed Numerical and Categorical spaces), a novel approach to address explanation using an anomaly detection algorithm, ADMNC, which provides accurate detections on mixed numerical and categorical input spaces. Our improved algorithm leverages the formulation of the ADMNC model to offer pre-hoc explainability based on CART (Classification and Regression Trees). The explanation is presented as a segmentation of the input data into homogeneous groups that can be described with a few variables, offering supervisors novel information for justifications. To prove scalability and interpretability, we list experimental results on real-world large datasets focusing on network intrusion detection domain.

Keywords: XAI; CART; anomaly detection; scalability; distributed computing; Apache Spark

1. Introduction

Anomaly Detection is an old discipline that has become relevant in situations in which datasets are huge and contain unexpected events carrying important information. These methods have found applications in fields such as network intrusion detection, and surveillance, among others. Several machine learning models are available [1,2], but despite being capable of offering very effective detection, most of these algorithms are unable to provide justifications about their outputs. The lack of explanation is one of the most important shortcomings of Machine Learning at present [3]. The European Union cites XAI (Explainable Artificial Intelligence) in its Ethics Guidelines for Trustworthy AI [4].

This work extends the ADMNC algorithm [5], an anomaly detection algorithm developed by our research group, with a new layer that opens the ADMNC black box by offering pre-hoc explainability. Regression decision trees are used to segment input data into homogeneous groups that can be described with a few variables. The objective is to provide a helpful and intuitive description of anomalous data, thus offering information to make informed decisions.

2. Methodology

The original ADMNC algorithm [5] is a method for large-scale offline learning to obtain a model of normal data that is then used to detect anomalies. The model used to obtain the pre-hoc explanation will consist of a grouping of the input patterns attending to their numerical variables. Clusters will be defined as the leaf nodes of a shallow decision tree [6]. Each pattern will be assigned its ADMNC estimator [5]. This estimator will then be approximated with a simple regression model, learned using the Apache Spark MLLib implementation of CART. Variance gives us an idea about how homogeneous the estimators for elements in a tree node are. Successive divisions turn nodes into more specific groups that contain similar elements. This balance between cluster homogeneity and explanation quality, given by the depth of each path, allows us to choose the level of detail for explanations.

We define the clustering $Cl(D)$ over dataset D as a set of m clusters Cl_i $\forall i \in [1, m]$ that contains every element in D. The *weighted variance* (WV) of a $Cl(D)$ is defined as:

$$WV(Cl(D)) = \frac{\sum\limits_{i \in 1..m} (\sigma^2{}_{Cl_i})|Cl_i|}{|D|}. \tag{1}$$

The weighted variance of a clustering measures how homogeneous its components are. This measure is complemented with another measure that indicates the number of input variables employed to characterize each cluster Cl_i. As a result, the *quality*, Q of a clustering is defined as:

$$Q(CL(D)) = -WV(Cl(D)) - \lambda \sum_{Cl_i \in Cl(D)} NV(Cl_i), \tag{2}$$

where $NV(Cl_i)$ represents the number of variables needed to describe cluster Cl_i and λ is a hyperparameter that allows the supervisor to balance the accuracy and interpretability [6] of the whole clustering. This quality measure is always negative and the goal of the algorithm is maximizing its value to approach 0. Maximizing this measure will ensure that the groups obtained are as homogeneous as possible and that they are explained using as few of the input variables as possible.

This method is carried out in two steps: (1) a full N level tree is built using the well-known CART algorithm. (2) This full tree is pruned to optimize the quality measure. Those node splits that decrease variance but also decrease quality are discarded, yielding a simpler tree that maximizes quality. The main features that lead data to be anomalous can be obtained as the path to anomalous clusters.

3. Experimental Results

To assess the validity of our approach, we considered two large datasets focusing on the network intrusion detection domain, KDDCup99 [5] and ISCXIDS 2012. For each resulting clustering, we measured its quality Q and weighted variance. We also included the number of clusters and the number of variables employed for both the full and pruned tree. These results are listed in Table 1. We set hyperparameter λ accordingly with pruning effort. This value can be modified by the supervisor, assigning more or less importance to interpretability in comparison to predictive power. Area under ROC (Receiver Operating Characteristic) curve is provided as fitness measure for anomaly detection, making five repetitions of each experiment. An example of explanatory tree is shown in Figure 1.

Table 1. Area under ROC curve (AUC) and explanatory tree metrics. Before pruning (Full, F) and after pruning (Pruned, P), considering hyperparameter λ, OV (Overall variance), Q (quality), WV (weighted variance), #Cl (number of clusters) and NV (number of variables to reach all clusters).

Dataset			AUC	Explanation				
Name	**OV**	λ	$(\mu \pm \sigma)$	**Tree**	Q	**WV**	**#Cl**	**NV**
ISCXIDS 2012	0.105	10^{-4}	0.919 ± 0.02	F	−0.062	0.048	29	142
				P	−0.051	0.049	7	25
KDDCup99 - FULL	0.049	10^{-3}	0.758 ± 0.05	F	−0.147	0.011	28	136
				P	−0.032	0.012	6	20
KDDCup99 - SMTP	2.846	10^{-3}	0.980 ± 0.01	F	−0.105	3.630×10^{-9}	22	105
				P	−0.005	6.632×10^{-6}	3	5
KDDCup99 - HTTP	0.843	10^{-3}	0.992 ± 0.01	F	−0.898	0.831	15	67
				P	−0.842	0.837	3	5
KDDCup99 - 10	2.454	10^{-3}	0.966 ± 0.02	F	−1.320	1.227	20	93
				P	−1.247	1.228	6	20

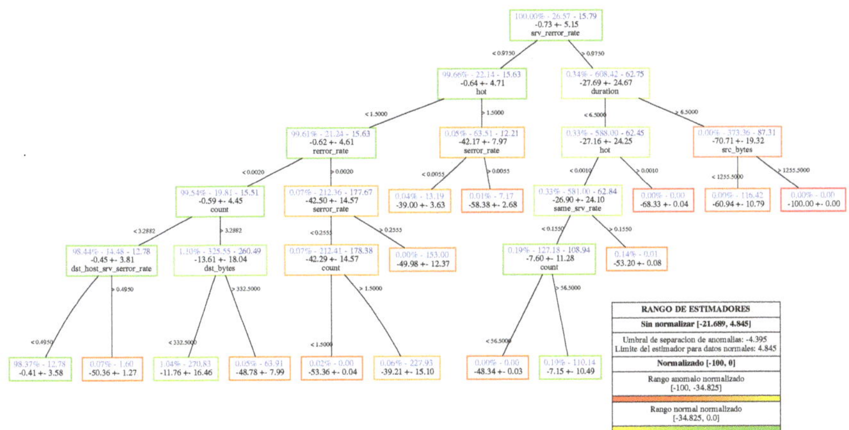

Figure 1. Explanatory tree after pruning ($\lambda = 10^{-3}$) using the KDDCup99-SMTP dataset. Named sequentially, reading from left to right, each node shows: the proportion of elements that it represents regarding the full dataset (shown in blue), overall variance (shown in blue), the weighted variance w.r.t children nodes (shown in dark blue) and mean and standard deviation for the subset of estimators. Further experimental results are given through supplementary materials reference.

4. Discussion and Conclusions

XAI is necessary to provide transparency to model predictions. It is a growing field of study that guarantees compliance with new European Union regulations. The proposed method allows us to examine differences between normal and anomalous data, potentially allowing the identification of generalization power, biases and formulation of hypothesis for abnormal data context.

In the future, we plan to add the categorical variables to the tree-based pre-hoc explanation. This will paint a more accurate picture of the input dataset. Another possible future research line is to improve explanations by introducing a previous dimensionality reduction step, as high dimensional data present redundant and irrelevant variables that produce bias and generalization errors.

Supplementary Materials: Pre-hoc regression trees are available online at https://www.dropbox.com/sh/m6lyn8zpss75sru/AADO_OFwzNwUTHD24vgJXhwma?dl=0

Funding: This research was partially funded by European Union ERDF funds, Ministerio de Ciencia e Innovación grant number PID2019-109238GB-C22, Xunta de Galicia through the accreditation of Centro Singular de Investigación 2016-2020, Ref. ED431G/01 and Grupos de Referencia Competitiva, Ref. GRC2014/035

Acknowledgments: We would like to thank CESGA for the use of their computing resources. Special recognition is given to the Spanish Ministerio de Educación for the predoctoral FPU funds, grant number FPU19/01457.

References

1. Liu, F.T.; Ting, K.; Zhou, Z. Isolation-Based Anomaly Detection. *TKDD* **2012**, *6*, 1–39.
2. Lu, Y.C.; Chen, F.; Wang, Y.; Lu, C.T. Discovering anomalies on mixed-type data using a generalized student-t based approach. *IEEE Trans. Knowl. Data Eng.* **2016**, *10*, 2582–2595.
3. Adadi, A.; Berrada, M. Peeking Inside the Black-Box: A Survey on Explainable Artificial Intelligence (XAI). *IEEE Access* **2018**, *6*, 52138–52160.
4. High Level Expert Group on Artificial Intelligence. Ethics Guidelines on Trustworthy Artificial Intelligence. 2019. Available online: https://ec.europa.eu/digital-single-market/en/news/ethics-guidelines-trustworthy-ai (accessed on 1 July 2020).

5. Eiras-Franco, C.; Martínez-Rego, D.; Guijarro-Berdiñas, B.; Alonso-Betanzos, A.; Bahamonde, A. Large Scale Anomaly Detection in Mixed Numerical and Categorical Input Spaces. *Inf. Sci.* **2019**, *487*, 115–127.
6. Eiras-Franco, C.; Guijarro-Berdiñas, B.; Alonso-Betanzos, A.; Bahamonde, A. A scalable decision-tree-based method to explain interactions in dyadic data. *Decis. Support Syst.* **2019**, *127*, 113141.

Proceedings

Network Anomaly Detection Using Machine Learning Techniques †

Julio J. Estévez-Pereira [1,*], Diego Fernández [1,2] and Francisco J. Novoa [1,2]

1 Department of Computer Science and Information Technologies, Faculty of Computer Science, Universidade da Coruña, Campus de Elviña, s/n, 15071 A Coruña, Spain; diego.fernandez@udc.es (D.F.); francisco.javier.novoa@udc.es (F.J.N.)
2 Research Center of Information and Communication Technologies (CITIC), Campus de Elviña, s/n, 15071 A Coruña, Spain
* Correspondence: julio.jairo.estevez.pereira@udc.es

† Presented at the 3rd XoveTIC Conference, A Coruña, Spain, 8–9 October 2020.

Published: 19 August 2020

Abstract: While traditional network security methods have been proven useful until now, the flexibility of machine learning techniques makes them a solid candidate in the current scene of our networks. In this paper, we assess how well the latter are capable of detecting security threats in a corporative network. To that end, we configure and compare several models to find the one which fits better with our needs. Furthermore, we distribute the computational load and storage so we can handle extensive volumes of data. The algorithms that we use to create our models, Random Forest, Naive Bayes, and Deep Neural Networks (DNN), are both divergent and tested in other papers in order to make our comparison richer. For the distribution phase, we operate with Apache Structured Streaming, PySpark, and MLlib. As for the results, it is relevant to mention that our dataset has been found to be effectively modelable with just a reduced number of features. Finally, given the outcomes obtained, we find this line of research encouraging and, therefore, this approach worth pursuing.

Keywords: machine leaning; IDS; network security; distributed computing; network flow

---

1. Introduction

A network anomaly can be defined as a variation of the regular behavior of the network. That includes both unfortunate unintended events, and deliberate attacks planned to compromise the network's availability. In both cases, it is essential to be able to detect those quickly so we can react in time [1].

In the past few years, machine learning has been slowly taking its place as an alternative to policies-based traditional intrusion detection systems, as it presents quite a few advantages. Machine learning techniques allow the development of non-parametric algorithms, adaptative to our network and its modifications, and portable across applications [1].

Although there is still a gap between the deployment of machine learning based intrusion detection systems and their predecessors due to different challenges [2], it is a reality that traditional detection methods fall behind when it comes to handle large-scale volumes of data, as their analysis processes are complex and time-consuming. Nevertheless, machine learning and big data tools and techniques can help us overcome these shortcomings [3].

In this paper, we go through different machine learning and big data alternatives, as we model the Intrusion Detection Evaluation Dataset (UNB ISCX IDS 2012) and deploy it so it supports extensive data processing. The solution we will be deploying will be eventually based on the best model and implemented on Apache Structured Streaming, using PySpark and MLlib. The storage will be also distributed with Hadoop. Furthermore, for performance monitoring, we will use Zabbix and JMX.

2. Results

2.1. Machine Learning Process

The phases that we went through in this project are those defined by CRISP-DM methodology: business understanding, data understanding, data preparation, modeling, evaluation, and deployment. In this subsection, we will be focusing on the modeling and evaluating phase, as our priority is to show the outcomes acquired. In the next subsection, we will address deployment related aspects.

2.1.1. Naive Bayes

For this project, we used a two-level Naïve Bayes model. First, we separated our binary features from our continuous ones. For the first group, we used Bernoulli Naïve Bayes, and for the latter Gaussian Naïve Bayes. After both were trained, we ensembled the probabilities given by each model for training a last Gaussian Naïve Bayes model. With this approach, we reached an 86.9% accuracy score in the ensembled model. As a relevant remark, we were reaching about a 96.9% accuracy in the ensembled model before taking out a couple of the correlated redundant features. These features happened to be quite descriptive, so, by letting them in, we were imposing a "positive bias" in the model, as the information encoded in those features was taken into account twice.

2.1.2. Random Forest

Due to the flexibility of Random Forest in terms of feature input, we could easily assess how well it behaves with different feature set approaches. We discovered that we could use a reduced set of original features and still get a 99.7% of accuracy on our dataset. After trying a wider range of features and applying hyperparametrization on the number of trees and the number of maximum features per tree split, we managed to raise that number to 99.8% with the configuration of 2000 trees and 15 maximun features per tree split.

2.1.3. Deep Neural Networks (DNN)

The DNN based model, as expected by the properties of the algorithm, was the one that entailed the largest amount of configuration of the three. Feature wise, we scaled the continuous variables, as DNN is quite sensitive to their values. For the initial architecture of the network, we started from the final disposition of layers and neurons described in Gabriel Fernandez's project [4]. Even so, after performing hyperparametrization, our architecture changed, ending with two hidden layers of 160 neurons each. The score was archived after all of the configurations were 99.6%.

2.2. Distribution

After all the algorithms were evaluated, we decided to prioritize the distribution of the one whose results were better, which is Random Forest. The next step consisted of reimplementing the model using PySpark and MLlib, which is the Spark's machine learning library. Then, we defined the right transformations to deliver the input dataset in the right format for Random Forest to process in a pipeline, and finally we trained the model. After that, we developed a PySpark script to submit to Spark in which we defined aspects such as the input and output streams of data, and the query we wanted to make over the streaming data.

We assessed the efficiency of the distribution strategy with one and four workers, comparing the metrics "inputRate" and "processingRate". We found out that the performances were very similar. In fact, the rates became slightly worse when using four workers. We think that this phenomenon is caused by the overhead introduced by Spark, as we use it both for the ingestion and processing of data.

3. Conclusions and Discussion

From a general point of view, we define the outcomes obtained as positive. They certainly align with both our hypothesis and other authors' ideas [1,3] about how machine learning in the network security domain can make a difference.

About the models, and concretely respecting Naïve Bayes, its simplicity and strong assumptions about feature independency [5] probably take a part in why it falls behind in terms of capability of prediction with respect to our other choices. On the other side, Random Forest overtook every other alternative in almost every aspect. With a default configuration and using only a few of the original features, it was already the best scoring algorithm. Concerning DNN, its outcomes came close to the ones of Random Forest. In addition, this is the algorithm that benefits the most from hyperparametrization.

In reference to the distribution outcomes, the results display an issue that needs to be addressed to successfully take advantage of one of the benefits that we have been claimed machine learning can offer, handling the processing of extensive volumes of data.

In future lines of research, we would perform similar experiments to other datasets as the results could be influenced to some extent by the peculiarities of the data. As for the models, we probably could somewhat improve the results for DNN if we explored more sophisticated feature engineering techniques, such as Feature Embedding. Finally, in view of the outcomes of the distribution phase, we would delegate the data ingestion to a separated solution, such as Kafka.

4. Materials and Methods

The dataset used for this project was the UNB ISCX IDS 2012, and the main tools that allowed us to work through all the phases of the machine learning process but the deployment were: Scikit-Learn v0.23.1, Tensorflow v2.1.0, Pandas v1.0.5, and Numpy v1.18.5. These supplied us with all the core functionality we needed for handling our dataset, modeling, and evaluating. However, we also made use of some complementary libraries. Concretely, for visualization, we employed Matplotlib v3.2.2 and Seaborn v0.10.1, and for the transformation of IP related features, Netaddr v0.8.0.

Our distributing architecture was based on Spark v3.0.0 and Hadoop v2.7.0. For developing applications on this architecture, we employed PySpark and MLlib v3.0.0. Finally, we delegated the monitoring and stream metrics visualization to Zabbix v5.0.0 and JMX.

Author Contributions: Conceptualization, D.F. and F.J.N.; methodology, J.J.E.-P., D.F., and F.J.N.; software, J.J.E.-P., D.F., and F.J.N.; validation, D.F. and F.J.N.; formal analysis, J.J.E.-P.; research, J.J.E.-P.; resources, J.J.E.-P.; data curation, J.J.E.-P.; writing—original draft preparation, J.J.E.-P.; writing—review and editing, J.J.E.-P., D.F. and F.J.N.; visualization, J.J.E.-P.; supervision, D.F. and F.J.N. All authors have read and agreed to the published version of the manuscript.

Conflicts of Interest: The authors declare no conflict of interest.

References

1. Ahmed, T.; Oreshkin, B.; Coates, M. Machine Learning Approaches to Network Anomaly Detection. In Proceedings of the Second Workshop on Tackling Computer Systems Problems with Machine Learning Techniques (SysML07), Cambridge, MA, USA, 10 April 2007.
2. Sommer, R.; Paxson, V. Outside the Closed World: On Using Machine Learning for Network Intrusion Detection. In Proceedings of the 2010 IEEE Symposium on Security and Privacy, Berkeley/Oakland, CA, USA, 16–19 May 2010; Volume 1.
3. Amrollahi, M.; Hadayeghparast, S.; Karimipour, H.; Derakhshan, F.; Srivastava, G. Enhancing Network Security Via Machine Learning: Opportunities and Challenges. In *Handbook Of Big Data Privacy*; Springer International Publishing: Cham, Switzerland, 2020; p. 169.

4. Fernandez, G. Deep Learning Approaches for Network Intrusion Detection. Ph.D. Thesis, The University of Texas at San Antonio, San Antonio, TX, USA, 2019; pp. 19, 20, 50, 51.
5. Panda, M.; Ranjan Patra, M. Network Intrusion Detection Using Naïve Bayes. *IJCSNS Int. J. Comput. Sci. Netw. Secur.* **2007**, *7*, 258–263.

Proceedings

Reflections about Learning Radiology inside the Multi-User Immersive Environment Second Life® during Confinement by Covid-19 †

Shaghayegh Ravaei *, Juan M. Alonso-Martinez, Alberto Jimenez-Zayas and Francisco Sendra-Portero *

Department of Radiology and Physical Medicine, School of Medicine, University of Málaga, 29071 Málaga, Spain; martin17alonso@gmail.com (J.M.A.-M.); albertojzayas@hotmail.es (A.J.-Z.)

* Correspondence: shaghayegh.ravaei@gmail.com (S.R.); sendra@uma.es (F.S.-P.)

† Presented at the 3rd XoveTIC Conference, A Coruña, Spain, 8–9 October 2020.

Published: 19 August 2020

Abstract: The multi-user immersive virtual environment Second Life® has been used to teach radiology to third-year medical students during confinement due to the current Covid-19 pandemic. In general, the students, who are digital natives nowadays, have found it easy to adapt to the use of the 3D platform. Although there have been some technical limitations, both students and teachers involved have rated the use of Second Life® during the confinement very highly.

Keywords: online learning; virtual worlds; medical students; undergraduate education; radiology

---

1. Introduction

The state of alarm decreed in Spain by Covid-19 determined that, as of 14 March 2020, the face-to-face activities at the university would be suspended and replaced by online training. In the Radiology core course of the third year of medical studies at the University of Malaga, the virtual immersive multi-user environment, Second Life®, has been used. The aim of this work is to reflect on the advantages, disadvantages and other peculiarities of this virtual platform in this exceptional situation.

2. Material and Methods

The following activities have been performed in Second Life® during the confinement: (i) 72 two-hour seminars for groups of 23–25 people, keeping the planned schedule and involving 8 teachers; (ii) 9 one-hour conferences with guest teachers; (iii) a virtual learning game, during 6 weeks, competing with students from other universities; (iv) an option to take the final oral practical exam. Several surveys have been performed with students and teachers about the ease of access to Second Life®, the cognitive load that they had to handle within the 3D environment or about the perception of the different activities carried out.

3. Results

There were 220 students enrolled in the Radiology course, 176 first-time students and 44 repeating students. The seminars were held in Second Life® by 157 students (71.4%), with 48 sporadic absences (0.4%) for various reasons. Forty-six first-time students (26.1%) voluntarily participated in the inter-university game. Fifty-four students attended the conferences given by guest professors. Sixteen students accepted to take the online oral exam in Second Life®. Five of the eight teachers of the subject and eight of the nine invited speakers had never used this platform as a teaching tool.

Eleven percent of students stated that their access to Second Life® was difficult or impossible. Table 1 and Figure 1 show some results from a survey about the seminars held in Second Life®.

Table 1. Evaluation of the radiology seminars given in Second Life®.

Assessed Item	Scores [1]
Connectivity to Second Life®	9.40 ± 1.06
The island environment and classrooms in Second Life®	9.43 ± 1.06
Audio quality in Second Life®	9.41 ± 1.00
The experience receiving seminars in Second Life®	9.52 ± 0.87
Teachers in Second Life®	9.30 ± 1.04
Interest in the content for third-year medical students	9.34 ± 0.93

[1] Sample: 84 out of 90 students answered these items. Data are mean ± standard deviation of scoring from 0 to 10 points.

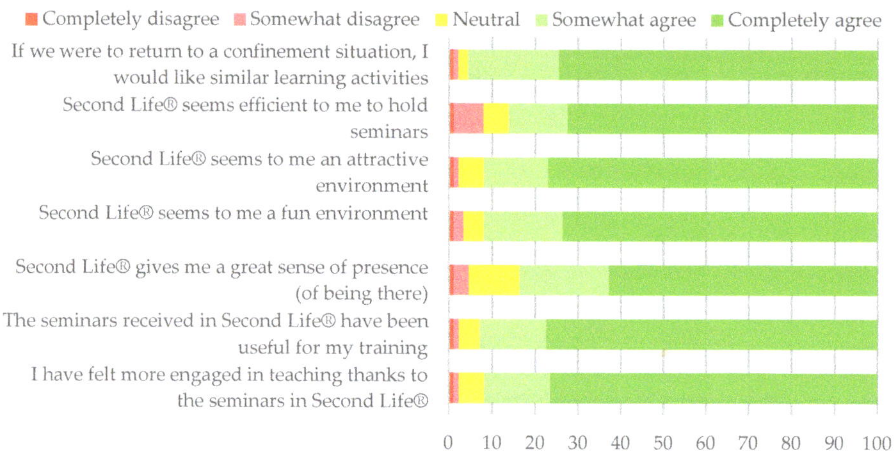

Figure 1. Bar chart of the percentages of answers in a 5-point Likert scale related to questions about radiology seminars held in Second Life®. Sample: 90 students (41% enrolled in the course).

4. Discussion

Second Life® is not a new educational resource in the subject of Radiology, as it has been used since the 2011–12 academic year without interruption [1–5]. This has allowed the virtual environment to be prepared for use in this critical situation. In fact, both the conferences with guest professors and the inter-university competitive game were activities scheduled before the state of alarm. These projects and the 3D platform were explained to all students at the beginning of the course, on February 14th. A month later, students suddenly found themselves in a situation of uncertainty, confined to their homes, not knowing how their training would be resolved. The practical seminars, which are essential in this subject, allowed for synchronous contact with their teachers, maintaining the scheduled calendar and providing a situation of linkage and continuity with the academic activity. The seminars with guest professors and the competitive game have provided complementary training options. The evaluation, which was usually done orally, has allowed Second Life® to be offered as a synchronous examination platform.

4.1. Benefits of Using Second Life to Teach during the Confination

- Ease of adaptation for current medical students (Generation Z) to interactive 3D platforms [6] facilitates the willingness to use Second Life®.

- The feeling of being present in the classroom, sitting next to classmates, is greater than in other 2D environments and helps to connect with teaching activity made impossible by confinement.
- Second Life® is a fun environment, emulating a role play between students and teachers, and served as a distraction to break the monotony of confinement.
- It facilitates student–teacher interaction, especially when personal contact is not possible.
- It also helps maintain peer relationships (it has been especially interesting in the learning game).
- It allows for the incorporation of permanent educative content to be used asynchronously 24/7.

4.2. *Drawbacks*

- Technical limitations due to low central processing unit (CPU), graphics card, or internet connection capacity may impede access to Second Life® or cause defects in the virtual world rendering.
- When the quality of the audio (voice input/output) of a user is limited, it hinders the correct development of synchronous activities. This becomes crucial when it happens to the teacher.
- Sometimes it is difficult to display web-based slideshows, especially when there are many interactions on the displaying screen. This may be minimized with proper pre-training.
- The limit of attendees to avoid access problems is around 45 avatars synchronously in the same place. This prevents large group lectures from being taught with Second Life®.

Although there have been some technical limitations, both students and teachers involved have rated the use of Second Life® during the confinement very highly.

Author Contributions: Conceptualization, S.R.; J.M.A.-M.; A.J.-Z. and F.S.-P.; methodology, S.R. and F.S.-P.; software, A.J.-Z. and F.S.-P.; validation, S.R.; J.M.A.-M.; A.J.-Z. and F.S.-P.; formal analysis, S.R. and F.S.-P.; investigation, S.R.; J.M.A.-M.; A.J.-Z. and F.S.-P.; resources, F.S.-P.; data curation, S.R.; J.M.A.-M.; A.J.-Z. and F.S.-P.; writing—original draft preparation, S.R. and F.S.-P.; writing—review and editing, S.R.; J.M.A.-M.; A.J.-Z. and F.S.-P.; visualization, S.R.; J.M.A.-M.; A.J.-Z. and F.S.-P.; supervision, F.S.-P.; project administration, F.S.-P.; funding acquisition, F.S.-P. All authors have read and agreed to the published version of the manuscript.

Funding: The Innovative Education Project #PIE19-217 of the University of Málaga partially supported this study. The maintenance cost of the Medical Master Island during this project was supported by the Andalusian Society of Radiology (Asociación de Radiólogos del Sur), a subsidiary of the Spanish Society of Medical Radiology (SERAM).

Acknowledgments: We want to express our appreciation to all the students and teachers who, due to the special circumstances of the pandemic, have agreed to participate in this teaching experience.

Conflicts of Interest: The authors declare no conflict of interest.

References

1. Sendra-Portero F.; Lorenzo-Alvarez R.; Pavia-Molina J. Teaching radiology in the "Second life" virtual world. *Diagn Imag. Eur.* **2018**, *34*, 43–45.
2. Lorenzo-Alvarez R.; Pavia-Molina J.; Sendra-Portero F. Exploring the potential of undergraduate radiology education in the virtual world Second Life with first-cycle and second-cycle medical students. *Acad. Radiol.* **2018**, *25*, 1087–1096.
3. Lorenzo-Alvarez R.; Rudolphi-Solero T.; Ruiz-Gomez M.J.; Sendra-Portero F. Game-based learning in virtual worlds: A multiuser online game for medical undergraduate radiology learning within Second Life. *Anat. Sci. Educ.* **2019**, in press.
4. Lorenzo-Alvarez R.; Ruiz-Gomez M.J.; Sendra-Portero F. Medical students' and family physicians' attitudes and perceptions toward radiology learning in the virtual world Second Life. *AJR Am. J. Roentgenol.* **2019**, *212*, 1295–1302.

5. Lorenzo-Alvarez R.; Rudolphi-Solero T.; Ruiz-Gomez M.J.; Sendra-Portero F. Medical student education for abdominal radiographs in a 3D virtual classroom versus traditional classroom: A randomized controlled trial. *AJR Am. J. Roentgenol.* **2019**, *213*, 644–650.
6. Eckleberry-Hunt J.; Lick D.; Hunt D. Is Medical Education Ready for Generation Z? *J. Grad. Med. Educ.* **2018**, *10*, 378–381.

Proceedings

Developing Open-Source Roguelike Games for Visually-Impaired Players by Using Low-Complexity NLP Techniques †

Luís Fernández-Núñez, Darío Penas, Jorge Viteri, Carlos Gómez-Rodríguez and Jesús Vilares *

CITIC, LyS Research Group, Department of Computer Science & Information Technologies, Facultade de Informática, Campus de Elviña, Universidade da Coruña, 15071 A Coruña, Spain; fernandezn.luis@gmail.com (L.F.-N.); dariosoyyo@gmail.com (D.P.); j.viteri.letamendia@udc.es (J.V.); carlos.gomez@udc.es (C.G.-R.)

* Correspondence: jesus.vilares@udc.es; Tel.: +34-881-01-1364

† Presented at the 3rd XoveTIC Conference, A Coruña, Spain, 8–9 October 2020.

Published: 19 August 2020

Abstract: The prominent graphic component of video games greatly limits the accessibility of this type of entertainment by visually impaired users. We make here an overview of the first games developed within an initiative for the development of roguelike games adapted to visually impaired players by using Natural Language Processing techniques. Our approach consists of integrating a multilingual module that automatically generates text descriptions of what is happening within the game. The user can then read such descriptions by means of a screen reader.

Keywords: Natural Language Generation; roguelikes; visually impaired users

1. Introduction

The offer of games accessible to users with severe visual impairments such as blindness is still limited both in genres and titles. This is mainly due to the prominent role played by graphics in games, thus limiting their accessibility. However, it may also happen when the graphic aspect of the game is secondary, as in the case of *roguelike* games. In these games, the player controls a hero who explores a dungeon complex fighting monsters, avoiding traps and discovering treasures. This genre has a great number of fans, both players and developers, who pay little attention to the graphic aspect on behalf of the game mechanics and the player experience. Actually, ASCII-based graphics are not rare even at present [1]. Nevertheless, little attention has been paid to their accessibility by blind players.

For several years, our research group has been offering to our students the chance of developing especially adapted open-source roguelike games as their Final Year Projects. These games should be accessible to both sighted and visually impaired players by applying low-complexity *Natural Language Processing* (NLP) techniques. To do this, these games offer, apart from the standard game mode, a so-called *descriptive mode* intended for visually impaired users. In this extra game mode, the classic graphic representation of the dungeon and its elements is replaced by natural language descriptions automatically generated by a multilingual module, which can be read using screen-reading software.

The system should be flexible and easily extensible and modifiable by third parties. Thus, other members of the game community could improve aspects of the basic game or add new languages to the description module. Our intention is that any person with basic programming skills and high-school linguistic skills should be able to extend the system, at least in part, to a new language. Additionally, our experience with low-resource languages suggested that the number and complexity of the linguistic resources to be used should be low. Therefore, they should be simple to obtain or build, if needed, since the free availability of these types of resources is often limited in many languages [2].

2. System Architecture

Figure 1 shows the general architecture of our games. The main difference with respect to regular roguelikes is the addition of a so-named *Text Description Engine* module. This new component takes as input the (current) world model and, together with the information provided by the game logic in response to the commands of the user, it generates a textual description of the dungeon and what is happening within the game. Such description is generated according to the linguistic resources available and the language selected. This new module follows the classic architecture of a Natural Language Generation (NLG) system, consisting of three stages [3]:

- *Content planning*. Determines the content and structure of the description.
- *Microplanning*. Selects the words and syntactic structure to be used to express such content.
- *Surface realisation*. Integrates all this information and transforms the abstract representation of the message into actual text to be presented to the user.

Seeking expressivity and variety, as stated before, especial attention has been paid to the microplanner. If the user is presented with monotonous descriptions that are repeated in the same terms again and again, he will soon get tired. To avoid this situation, we take advantage of the so-called *linguistic variation* (a.k.a. *linguistic flexibility*): the ability of our languages to express the same message in very different ways (and vice versa) [4].

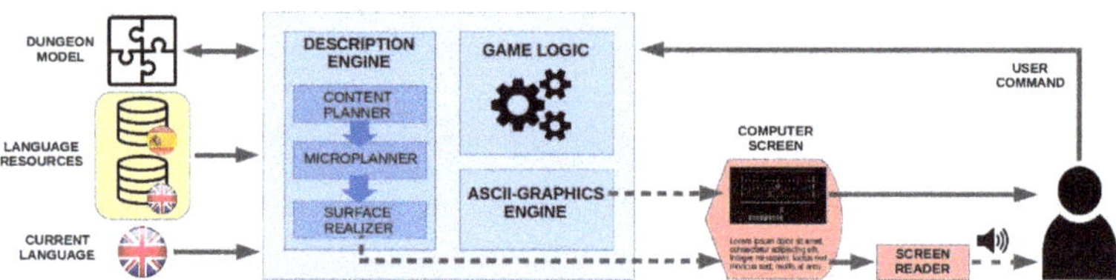

Figure 1. General architecture of the system.

3. Linguistic Processing

The techniques applied when implementing the generation module involve different levels of linguistic processing, from the lexical to the pragmatic level, without losing sight of multilingualism.

At the *lexical and morphological levels*, it is necessary to provide the system with a *vocabulary* containing every element, action and situation that may occur or appear within the game. For every term, this dictionary also includes its corresponding morphosyntactic information (e.g., word category, part-of-speech tag, etc.) together with its inflectional variants due to changes in gender, number, person, etc. Moreover, every term must be conveniently indexed with an external identifier through which it will be referenced in those program entities or processes involving it. For each language available in the game, a separate vocabulary must be created, with all the translations of a given term sharing the same external identifier. These dictionaries were implemented as JSON files because of their simplicity and flexibility.

Regarding the *syntactic level*, once the description module has retrieved those vocabulary terms corresponding to a gameplay event, they are arranged according to a *generation grammar* to form a meaningful message describing the event. This grammar is structured into *subgrammars* according to the syntactic structures generated (e.g., noun phrases) and the different contexts in which they are used (e.g., combat). Again, subgrammars and rules are indexed using an external identifier to link them with the events they describe, independently of the specific language being used at a given time.

These grammars are defined using context-free rules and kept in the form of JSON files using a format inspired in feature structure-based grammars [5].

With respect to the *semantic Level*, we have experimented with an alternative vocabulary structure where terms are not selected and organized individually, but grouped forming *synsets* (synonym sets) instead. Again, an external identifier has been used to identify and link the entries (i.e., synsets) between languages.

Finally, at the *discourse and pragmatic levels*, some features were successfully tested. One of them consisted on changing the adjectives used to describe a character taking into account its current state (e.g., wounded or hungry), thus introducing a subjective point of view. Another feature was persistence over time; for example, after having combat in a room, subsequent descriptions of that room should reflect damages and signs of a fight.

4. Results

Until now, three games were developed within our initiative for developing open-source roguelike games accessible for visually impaired users by using low-complexity NLP techniques:

1. *The Inner Eye*, by Luis Fernández-Núñez;
2. *The Accessible Dungeon*, by Darío Penas;
3. and *Hamsun's Amulet*, by Jorge Viteri.

all af them available at our website. At this point, the languages that were addressed are: English, Spanish, Galician and Dutch. Preliminary work has been made for Japanese, too. Further details about the solutions applied for the implementation of these games can be found in our previous work [6].

From a social point of view, our proposal favors integration, since both sighted and visually impaired users can play the same game and, thus, share common experiences. By helping draw students' attention to accessibility concerns, students will be aware of them when they go on to become software developers. Finally, from an academic point of view, this type of final-year projects allow us to introduce the student to NLP in a practical and engaging way.

Author Contributions: This work derives from the Bachelor's and Masther's theses of L.F.-N., D.P. and J.V. (Jorge Viteri), all of them co-supervised by J.V. (Jesús Vilares) and C.G.-R. All authors have read and agreed to the published version of the manuscript.

Funding: The work of Profs. Vilares and Gómez-Rodríguez was partially funded by MINECO (through project ANSWER-ASAP TIN2017-85160-C2-1-R) and Xunta de Galicia (through grants ED431B 2017/01, ED431G/01 and ED431G 2019/01).

Acknowledgments: We want to express our gratitude to the roguelike and blind gaming community, especially to members of *Tiglo-juegos* Google group, and to Juan Carlos Buño-Suárez, trainer from the National Organization of Spanish Blind People (ONCE), for their help in sharing their needs, suggesting improvements and betatesting.

Conflicts of Interest: The authors declare no conflict of interest.

References

1. Craddock, D.L. *Dungeon Hacks: How Nethack, Angband, and other Roguelikes Changed the Course of Video Games*; Press Start Press: Canton, OH, USA, 2015.
2. Rehm, G.; Uszkoreit, H. (Eds.) *META-NET White Paper Series*; Springer: Berlin, Germany, 2011. Available online: http://www.meta-net.eu/whitepapers (accessed on 18 August 2020).
3. Reiter, E.; Dale, R. *Building Natural Language Generation Systems*; Cambridge University Press: Cambridge, UK, 2000.
4. Arampatzis, A.; van der Weide, T.P.; van Bommel, P.; Koster, C. Linguistically-motivated Information Retrieval. In *Encyclopedia of Library and Information Science*; Marcel Dekker: New York, NY, USA, 2000; Volume 69, pp. 201–222.
5. Carpenter, B. *The Logic of Typed Feature Structures*; Number 32 in Cambridge Tracts in Theoretical Computer Science; Cambridge University Press: Cambridge, UK, 1992.

6. Vilares, J.; Gómez-Rodríguez, C.; Fernández-Núñez, L.; Penas, D.; Viteri, J. Bringing Roguelikes to Visually-Impaired Players by Using NLP. In Proceedings of the LREC 2020 Workshop Games and Natural Language Processing, Marseille, France, 11 May 2020; Lukin, S.M., Ed.; European Language Resources Association (ELRA): Paris, France, 2020; pp. 59–67. Available online: https://lrec2020.lrec-conf.org/media/proceedings/Workshops/Books/Games-NLPbook.pdf (accessed on 18 August 2020).

Proceedings

On the Effectiveness of Convolutional Autoencoders on Image-Based Personalized Recommender Systems †

Eva Blanco-Mallo [1,*], Beatriz Remeseiro [2], Verónica Bolón-Canedo [1] and Amparo Alonso-Betanzos [1]

1 Campus de Elviña s/n, Universidade da Coruña, CITIC, 15071 A Coruña, Spain; vbolon@udc.es (V.B.-C.); amparo.alonso.betanzos@udc.es (A.A.-B.)
2 Campus de Gijón s/n, Universidad de Oviedo, 33203 Gijón, Spain; bremeseiro@uniovi.es
* Correspondence: eva.blanco@udc.es
† Presented at the 3rd XoveTIC Conference, A Coruña, Spain, 8–9 October 2020.

Published: 19 August 2020

Abstract: Over the years, the success of recommender systems has become remarkable. Due to the massive arrival of options that a consumer can have at his/her reach, a collaborative environment was generated, where users from all over the world seek and share their opinions based on all types of products. Specifically, millions of images tagged with users' tastes are available on the web. Therefore, the application of deep learning techniques to solve these types of tasks has become a key issue, and there is a growing interest in the use of images to solve them, particularly through feature extraction. This work explores the potential of using only images as sources of information for modeling users' tastes and proposes a method to provide gastronomic recommendations based on them. To achieve this, we focus on the pre-processing and encoding of the images, proposing the use of a pre-trained convolutional autoencoder as feature extractor. We compare our method with the standard approach of using convolutional neural networks and study the effect of applying transfer learning, reflecting how it is better to use only the specific knowledge of the target domain in this case, even if fewer examples are available.

Keywords: personalized recommendation; image-based recommendation system; feature extraction; convolutional autoencoder; convolutional neural network; data augmentation

---

1. Introduction

With the advent of e-commerce, social networks dedicated to sharing product reviews began to become popular, leading to the integration of different personalized recommender systems (RS) with great success on various on-line platforms. Starting from the premise that a picture is worth more than a thousand words [1], our goal is to make use of the data available in TripAdvisor to build a personalized gastronomic image-based RS. One of the first attempts to use visual information in personalized RS was proposed by He et al. [2], using a convolutional neural network (CNN) to extract the deep features of product images that are then processed through matrix factorization. As for the use of TripAdvisor data, Zhang et al. [3,4] also use them for recommendation purposes, but do not consider the visual information at all. A more related approach to ours is the one presented by Díez et al. [1], who also use TripAdvisor images but to determine authorship of the images, aiming at providing explainable recommendations. To the best of our knowledge, our work is the first one that studies the effect of using only images to characterize user tastes in a personalized recommendation system. Unlike existing approaches, we propose to use a convolutional autoencoder (CAE) as a feature extractor, because (1) it works better than standard approaches based on pre-trained CNN, and (2) it is less computationally expensive.

2. Materials and Methods

The problem addressed in this manuscript is defined as a binary classification task in which we have some triads with either one of two labels, $(u, r, i) \rightsquigarrow 0|1$, where u is a user who took the image i of a restaurant r he/she visited. As for the two labels, 0 means that the user u does not like the restaurant r reflected in i, while 1 stands for the opposite. A network that learns on triads of users and photos (u, r, i) is proposed to provide a personalized image-based recommendation. Users and restaurants are represented by a one-hot codification and then mapped into 512-dimensional embeddings, while photos are codified by a CAE. The CAE, based on the one proposed by Chollet [5], was trained using the entire set of images available, and then used to encode the images used at the model input. This code is processed by a fully connected (FC) layer, and then concatenated with the one-hot embeddings, resulting in a single vector of 1536 features. Next, a series of FC, ReLU and dropout layers follow, ending with the sigmoid activation function that generates a probability value, where 0 means that the user u does not like the restaurant r depicted by i, and 1 that he/she likes it.

The data used for experimentation purposes were collected in 2018–2019 from TripAdvisor reviews of restaurants in three cities of different sizes [1]: Santiago de Compostela (SGC), Barcelona (BCN) and New York (NYC). The original dataset was divided to obtain the training and test sets, following this procedure: for each user, reviews are separated into positive and negative, with 1 review (if there is more than one) for the test set and $N - 1$ for the training set; once this procedure is completed, if there is any review in the test set that belongs to a restaurant that is not included in the train set, then all its images are moved to the train set. The idea is to guarantee that all the users and restaurants evaluated in the test set are also in the train set. To select the hyper-parameters of the proposed architecture, a validation set was obtained applying the above procedure to the training set. Due to the high imbalance of the data, oversampling is applied on the minority class until an approximate balancing ratio of 1 is reached.

3. Results

All the experiments were performed on a computer equipped with a GeForce Titan XP 12GB GPU from NVIDIA, an Intel Core i7-4790 CPU @ 3.60GHz x 8, and 16 GiB memory. The implementation of the model and baseline methods is in Keras (https://keras.io/), using the Adam as optimizer and the HeUniform for weight initialization. The training process was carried out by setting a batch size of 32, a patience of 12 and a maximum of 100 epochs. The outputs were monitored by using the balanced score (B-score) metric, which represents the harmonic mean of sensitivity and specificity: B-score $= 2 *$ ((sensitivity * specificity) / (sensitivity + specificity)). The effectiveness of the CAE as a feature extractor is contrasted with a CNN, considered a benchmark in supervised image classification. Specifically, ResNet50 [6] with weights pre-trained on ImageNet is used, with and without parameter fine-tuning. In an RS, not only is it important to recommend those items that the user likes, but also not to recommend those items that the user does not like. Thus, we focus on sensitivity and specificity metrics and propose to use the B-score previously defined.

In light of the results shown in Table 1, the best performance is achieved by the CAE, specifically regarding specificity. The only city where the ResNet50 achieves a higher specificity is SGC and, even so, the balance between the classification of both classes is still lower than that obtained with the CAE. In general, terms, the worst scenario occurs when the ResNet50 is used without parameter fine-tuning, since it was trained on ImageNet and it is not adjusted to the problem at hand. Therefore, it is important to emphasize that the use of transfer learning was not as useful as intended and better results are obtained by using only specific knowledge of the target domain. In relation to the time per epoch, the CAE stands out without a doubt, requiring less than half the time that ResNet50 does and even less than nine times less if parameter fine-tuning is applied. Hence, our approach is more cost-effective not only in terms of performance, but also of processing time.

Table 1. Comparison of results using three different techniques for the image encoding step: using a CAE, a CNN pre-trained on ImageNet and a fine-tuned CNN (CNN_FT).

	Sensitivity			Specificity			B-Score		
	CAE	CNN	CNN_FT	CAE	CNN	CNN_FT	CAE	CNN	CNN_FT
NYC	0.7454	0.7261	0.7933	0.7594	0.7071	0.6724	0.7523	0.7165	0.7279
BCN	0.6175	0.8546	0.6738	0.8006	0.4480	0.6176	0.6972	0.5878	0.6445
SGC	0.7629	0.8453	0.7318	0.7905	0.6047	0.8181	0.7765	0.7050	0.7725

4. Conclusions

In this work, we explore the potential of modeling both users and restaurants using images as single source, demonstrating that the use of CAEs rather than CNNs is more convenient, not only considering performance but also resources. Taking into account that the high imbalance of classes and the few examples per user are characteristics in the context of the recommendations, it is interesting to further investigate possible transfer learning approaches, other than those based on parameter transfer, that may be more suitable in these scenarios. Our future research involves the integration of the proposed method into an RS that takes into account additional information, and its application to other RS that already deal with images.

Funding: This research has been financially supported in part by European Union FEDER funds, by the Spanish Ministerio de Economía y Competitividad (research project PID2019-109238GB), by the Consellería de Industria of the Xunta de Galicia (research project GRC2014/035), and by the Principado de Asturias Regional Government (research project IDI-2018-000176). CITIC as a Research Center of the Galician University System is financed by the Consellería de Educación, Universidades e Formación Profesional (Xunta de Galicia) through the ERDF (80%), Operational Programme ERDF Galicia 2014–2020 and the remaining 20% by the Secretaria Xeral de Universidades (ref. ED431G 2019/01).

Acknowledgments: Thanks to NVIDIA Corporation for donating the Titan Xp GPUs used in this research.

Conflicts of Interest: The authors declare no conflict of interest.

References

1. Díez, J.; Pérez-Núñez, P.; Luaces, O.; Remeseiro, B.; Bahamonde, A. Towards explainable personalized recommendations by learning from users' photos. *Inf. Sci.* **2020**, *520*, 416–430.
2. He, R.; McAuley, J. VBPR: visual bayesian personalized ranking from implicit feedback. In Prcoceedings of the Thirtieth AAAI Conference on Artificial Intelligence, Phoenix, AZ, USA, 12–17 February 2016; pp. 144–150.
3. Zhang, H.; Ji, P.; Wang, J.; Chen, X. A novel decision support model for satisfactory restaurants utilizing social information: A case study of TripAdvisor.com. *Tour. Manag.* **2017**, *59*, 281–297.
4. Zhang, C.; Zhang, H.; Wang, J. Personalized restaurant recommendation method combining group correlations and customer preferences. *Inf. Sci.* **2018**, *454*, 128–143.
5. Chollet, F. Building Autoencoders in Keras. *The Keras Blog*; **2016**.
6. He, K.; Zhang, X.; Ren, S.; Sun, J. Deep residual learning for image recognition. In Prcoceedings of the IEEE Conference on Computer Vision and Pattern Recognition, Las Vegas, NV, USA, 30 June 2016; pp. 770–778.

Proceedings

Application of Artificial Neural Networks for the Monitoring of Episodes of High Toxicity by DSP in Mussel Production Areas in Galicia †

Andrés Molares *, Enrique Fernandez-Blanco and Daniel Rivero

CITIC, Faculty of Computer Science, Campus de Elviña, University of A Coruña, 15006 A Coruña, Spain; enrique.fernandez@udc.es (E.F.-B.); daniel.rivero@udc.es (D.R.)

* Correspondence: andres.molares@udc.es

† Presented at the 3rd XoveTIC Conference, A Coruña, Spain, 8–9 October 2020.

Published: 19 August 2020

Abstract: This study seeks to support, through the use of Artificial Neural Networks (ANN), the decision to perform closings after days without sampling in the Vigo estuary. The opening and closing of the mussel production areas are based on the toxicity analysis of this bivalve's meat. Sometimes it is not possible to obtain the necessary data for effective closing. If there is evidence of an increase in toxicity levels, "Precautionary Closings" on mussel extraction is done. A small error in the forecast of the state of the areas could mean serious losses for the mussel industry and a huge risk for public health. Unlike in previous studies, this study aims to manage the state of the mussel production areas, whilst the others focused on predicting the harmful algae blooms. Having achieved test sensitivity values of 67.40% and test accuracy of 83.00%, these results may lead to new research that involves obtaining more accurate models that can be integrated into a support system.

Keywords: machine learning; harmful algae blooms; artificial neural networks

---

1. Introduction

Since 1995, a governmental monitoring program has managed the mussel production areas in Galicia. The creation of this program was necessary because of the high frequency of a phenomenon called Harmful Algal Blooms (HAB), which implies a temporary cessation in the extraction and commercialization of the mussels. The HAB are episodes of a high concentration of algae potentially toxic to humans through mussel consumption. In the Vigo estuary, the most common toxin-producing species are the DSP type, such as the *Dinophysis acuminata* dinoflagellate [1].

A weak point of this process is the absence of sampling during weekends or inclement weather, which sometimes makes it impossible to collect the data to support an effective closing. If there is an indication of an increase in levels of toxicity, the competent authority is legally empowered to proceed to the "Precautionary Closings" on the extraction of bivalve molluscs. Nowadays, the performance of this kind of closing is based on the expertise of government agents. A mathematical model to support the making of these decisions could help experts in complex situations that may cause errors in the decisions made.

Although the previously described situation is focused on the Galician Coast, this scenario is replicated in other major producers around the globe. That is why other works have tried to monitor the HAB episodes using different techniques. To date, those previous works have focused their efforts on predicting biomarkers, such as the concentration of toxic phytoplankton or chlorophyll "a". These studies, although of high scientific interest, do not give concrete support when it comes to monitoring the state of the production areas. The toxicity levels present in mussel meat depend on additional factors, such as retention of toxicity or the relationship between toxic versus non-toxic phytoplankton present in the medium. These factors, and some others, will be considered in conducting this study to achieve a more practical approach. To do that, a classifier based on Artificial Neural Networks (ANN) [2] is going to be defined to assess the state of the production areas affected by DSP-type toxins, on days with the absence of previous samplings.

2. Results

A summary of the obtained results can be seen in Figure 1. Each row of this table represents a tested model, where the first and second columns define the filters of characteristics used. These filters are represented with the value of the quartile selected for training. The third column shows the architecture of the networks by showing the number of hidden neurons per layer. The fourth column contains the p-value obtained from a Tukey–Kramer paired analysis, after previously performing a ANOVA analysis. Additionally, the remaining columns show the performance measures obtained in the test, that is, the average accuracy, average sensitivity, average kappa coefficient, minimum accuracy, minimum sensitivity, and minimum kappa coefficient.

Table ordered by sensibility									
Correlation filter quartile	Random Forest filter quartile	Neurons per layer	p	Accuracy $\bar{x}$	Sensibility $\bar{x}$	Kappa index $\bar{x}$	Accuracy Min	Sensibility Min	Kappa index Min
-	-	14		83.00%	67.40%	0.591	72.00%	36.00%	0.31
-	-	[10, 20]	1	83.50%	66.40%	0.599	72.00%	32.00%	0.29
-	-	[10, 10]	1	83.30%	66.20%	0.597	71.00%	30.00%	0.27
-	-	8	1	82.50%	63.30%	0.566	71.00%	30.00%	0.26
25	-	14	1	79.30%	60.00%	0.508	63.00%	27.00%	0.18

Figure 1. Statistics of the best models applied in the Vigo. A production area, ordered by descending sensibility. Access to the full table can be obtained here: https://github.com/AndresMolares/XoveTIC2020.git.

3. Discussion

After carrying out the study, it can be seen how ANN works better with a large number of characteristics to solve this problem. Although the works carried out to date obtain good results when making HAB predictions on the Galician coast (an overall accuracy between 78.53–82.18% using vector support machines to predict HAB of *Pseudo-nitzschia spp.* [3]), the control of the state of the production areas is conditioned by other external factors, so the definition of the problem changes. As this is the first study that seeks to provide support when estimating the state of the mussel production areas affected by DSP-type toxins, the results are promising (accuracy of 83%). However, to develop models that are precise enough to be integrated into support tools, it would be necessary to develop models with better sensitivity and accuracy values. For this, new machine learning algorithms could be studied, as well as a more exhaustive exploration of the hyperparameter space of the ANN.

4. Materials and Methods

Data from different sources were combined to create the dataset used in this experiment. Those sources contained different values sampled weekly between 2004 and 2018. The result is a dataset with the following variables: seasonality, concentration of chlorophyll "a", *Dinophysis acuminata*, ammonium, phosphate, nitrite, nitrate, water temperature, oxygen in water, salinity, solar irradiation, upwelling index, and the previous state of the production area. These data have been provided by the INTECMAR [4], METEOGALICIA [5], and IEO [6].

From raw data, two types of filtering were applied to choose the most significant features: (a) Applying a correlation matrix of the input variables with the state of the zone (variable objective); and (b) using a Random Forest algorithm as a discriminator. The Random Forest algorithm calculates the importance of a variable, taking into account how much the prediction error increases when the data for that variable is permuted, while all others remain unchanged.

With these methods, blocks of characteristics of quartiles 25, 50, and 75 were obtained. Different experiments have been defined based on the application of each one, another, both, or none of the previously mentioned filtering methods and the architecture of the ANN model. To ensure reliable results, the tests were performed with a 10-fold cross-validation strategy, which was repeated 50 times for each combination of filter methods and classification techniques. The reason for repeating this process is due to the non-deterministic nature of the backpropagation [7] used to train the ANN of each fold. To perform the training, each ANN model was set to use Dense Hidden layers, with the ADAM algorithm as an optimizer and binary cross-entropy as the loss function. Finally, the transfer function of the output layer is a sigmoid function in all cases exposed, while on the contrary, the activation function of the hidden layers is a Relu function.

With that configuration and data in common and having the same input and desired output, five models were tested by changing the number of hidden layers and the number of elements in these. More specifically, the trained and tested models were: One hidden layer with 2, 8, or 14 neurons; and two hidden layers with 10 neurons each and 10 and 20 neurons, respectively.

Author Contributions: Conceptualization, A.M.; methodology, E.F.-B.; software, A.M.; validation, A.M., E.F.-B. and D.R.; formal analysis, A.M. and D.R.; investigation, A.M.; resources, A.M.; data curation, A.M.; writing–original draft preparation, A.M.; writing–review and editing, E.F.-B.; visualization, A.M.; supervision, D.R.; project administration, D.R.; funding acquisition, A.M., E.F.-B. and D.R. All authors have read and agreed to the published version of the manuscript.

Funding: This work has been partially supported by different grants and projects from the Xunta de Galicia [ED431D 2017/23; ED431D 2017/16; ED431G/01; ED431C 2018/49].

Acknowledgments: The authors want to acknowledge the support from INTECMAR [4], who have provide the data for this work and CESGA [8], who allows to conduct the tests on their installations.

Conflicts of Interest: The authors declare no conflict of interest. The funders had no role in the design of the study; in the collection, analyses, or interpretation of data; in the writing of the manuscript, or in the decision to publish the results.

Abbreviations

The following abbreviations are used in this manuscript:

ANN	Artificial Neural Network
DSP	Diarrheal Shellfish Poisoning

References

1. Vilas, F.; Rey, D.; Rubio Armesto, B.; Bernabéu, A.; Méndez, G.; Durán, R.; Mohamed, K.; Rosón, G.; Cabanas, J.; Pérez, F.F.; et al. *La Ría de Vigo: una aproximación integral al ecosistema marino de la Ría de Vigo*; Instituto de Estudios Vigueses: Vigo, Spian, 2008.
2. McCulloch, W.S.; Pitts, W. A logical calculus of the ideas immanent in nervous activity. *Bull. Math. Biophys.* **1943**, *5*, 115–133.
3. Vilas, L.G.; Spyrakos, E.; Palenzuela, J.M.T.; Pazos, Y. Support Vector Machine-based method for predicting Pseudo-nitzschia spp. blooms in coastal waters (Galician rias, NW Spain). *Prog. Oceanogr.* **2014**, *124*, 66–77.
4. Instituto Tecnolóxico Para o Control do Medio Mariño de Galicia web page. Available online: http://www.intecmar.gal/intecmar/default.aspx (accessed on 17 January 2020).
5. METEOGALICIA Web Page. Available online: https://www.meteogalicia.gal/observacion/estacionshistorico/historico.action?idEst=14001 (accessed on 17 February 2020).
6. Marnaraia. Available online: http://www.indicedeafloramiento.ieo.es/afloramiento.html (accessed on 17 January 2020).

7. Hecht-Nielsen, R. Theory of the backpropagation neural network. In *Neural Networks for Perception*; Elsevier: Amsterdam, The Netherlands, 1992; pp. 65–93.
8. CESGA Web Page. Available online: https://www.cesga.es/ (accessed on 20 February 2020).

Proceedings

Numerical Simulation of a Nonlinear Problem Arising in Heat Transfer and Magnetostatics †

María González * and Hiram Varela

M2NICA Research Group, CITIC, Universidade da Coruña, 15071 A Coruña, Spain; hiram.varela@udc.es
* Correspondence: maria.gonzalez.taboada@udc.es
† Presented at the 3rd XoveTIC Conference, A Coruña, Spain, 8–9 October 2020.

Published: 19 August 2020

Abstract: We present a numerical model that comprises a nonlinear partial differential equation. We apply an adaptive stabilised mixed finite element method based on an a posteriori error indicator derived for this particular problem. We describe the numerical algorithm and some numerical results.

Keywords: nonlinear boundary value problem; finite element analysis; adaptivity

---

1. Introduction

We consider the mathematical model from [1]. The model comprises a second order nonlinear elliptic partial differential equation over a bounded domain Ω in $\mathbb{R}^2$, provided with Dirichlet boundary conditions over the contour Γ:

$$\left\{ \begin{array}{rcl} -\nabla\cdot(k(\cdot,|\nabla u|)\nabla u) & = & f \quad \text{in}\,\Omega, \\ u & = & g \quad \text{on}\,\Gamma. \end{array} \right. \tag{1}$$

The unknown u can be temperature, magnetic potential, etc., $|\cdot|$ denotes the Euclidean norm and ∇ and $\nabla\cdot$ denote, respectively, the usual gradient and divergence operators. The functions k, f and g are data, and we assume that $k:\Omega\times[0,+\infty)\to\mathbb{R}$ satisfies the following bound, for two positive constants k_1 and k_2: $k_1 \le k(\mathbf{x},s)+s\frac{\partial}{\partial s}k(\mathbf{x},s)\le k_2, \forall(\mathbf{x},s)\in\Omega\times[0,+\infty)$.

Problem (1) allows to predict a wide range of physical phenomena, such as magnetostatics and heat transfer. Due to the nonlinearity, we need to resort to numerical methods. We apply an adaptive stabilized mixed finite element method and present some results from our numerical experiments.

2. Adaptive Augmented Mixed Finite Element Method

We introduce two additional unknowns, $\mathbf{t}=\nabla u$ and $\sigma = k(\cdot,|\mathbf{t}|)\mathbf{t}$, and define the function space $X := [L^2(\Omega)]^2 \times H(\nabla\cdot,\Omega)\times L^2(\Omega)$, where $L^2(\Omega)$ is the space of square integrable functions and $H(\nabla\cdot,\Omega) := \{\mathbf{v}\in[L^2(\Omega)]^2; \nabla\cdot\mathbf{v}\in L^2(\Omega)\}$. In order to approximate the triplet $(\mathbf{t},\sigma,u)$, the domain is divided into triangular cells that conform a mesh (see Figure 1a). In each of those cells an approximation of the unknowns is calculated. The corresponding function space is $X_h := \mathcal{P}_0\times\mathcal{RT}_0\times\mathcal{P}_1$, with $\mathcal{P}_k$ being the space of polynomials of degree $\le k$ and $\mathcal{RT}_0$ is the lowest order Raviart Thomas space. We consider the following discrete problem: find $(\mathbf{t}_h,\boldsymbol{\sigma}_h,u_h)\in X_h$ such that

$$A((\mathbf{t}_h,\boldsymbol{\sigma}_h,u_h),(\mathbf{s}_h,\boldsymbol{\tau}_h,v_h)) = F(\mathbf{s}_h,\boldsymbol{\tau}_h,v_h), \quad \forall(\mathbf{s}_h,\boldsymbol{\tau}_h,v_h)\in X_h \tag{2}$$

where

$$A((\mathbf{t},\boldsymbol{\sigma},u),(\mathbf{s},\boldsymbol{\tau},v)) := \int_\Omega k(\cdot,|\mathbf{t}|)\mathbf{t}\cdot\mathbf{s} - \int_\Omega \boldsymbol{\sigma}\cdot\mathbf{s} + \int_\Omega \boldsymbol{\tau}\cdot\mathbf{t} + \int_\Omega u\nabla\cdot\boldsymbol{\tau} - \int_\Omega v\nabla\cdot\boldsymbol{\sigma} + \\ \kappa_1\int_\Omega(\boldsymbol{\sigma}-k(\cdot,|\mathbf{t}|)\mathbf{t})\cdot\boldsymbol{\tau} + \kappa_2\int_\Omega \nabla\cdot\boldsymbol{\sigma}\,\nabla\cdot\boldsymbol{\tau} + \kappa_3\int_\Omega(\nabla u-\mathbf{t})\cdot(\nabla v+\mathbf{s}) + \kappa_4\int_\Gamma u\,v \tag{3}$$

$$F(\mathbf{s},\boldsymbol{\tau},v) := \int_\Gamma \boldsymbol{\tau}\cdot\mathbf{n}\,g + \int_\Omega f\,v - \kappa_2\int_\Omega f\,\nabla\cdot\boldsymbol{\tau} + \kappa_4\int_\Gamma g\,v \tag{4}$$

with κ_i being positive constants. The nonlinear form $A(\cdot,\cdot)$ is strongly monotone, Lipschitz-continuous and bounded for appropriate values of the parameters κ_i so that problem (2) has a unique solution.

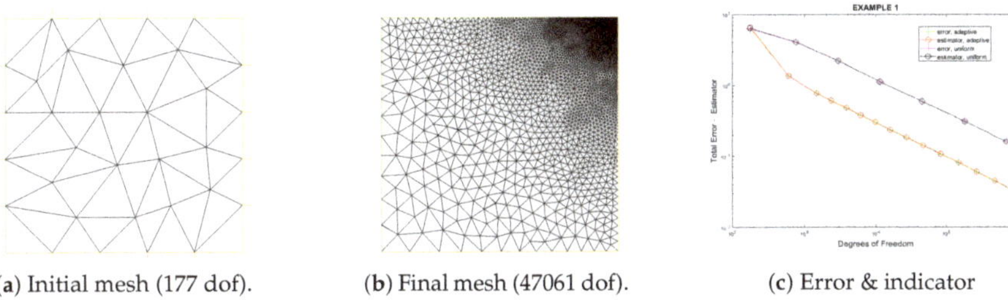

(**a**) Initial mesh (177 dof). (**b**) Final mesh (47061 dof). (**c**) Error & indicator

Figure 1. Some meshes, and error and indicator for the uniform and adaptive refinements vs. dof.

We consider an adaptive algorithm, which estimates the local error in each cell, and refines the mesh over those regions with the highest error. On each triangle, we introduce the following novel error indicator, which is reliable and locally efficient [2]:

$$\theta_T^2 := \|\nabla u_h - \mathbf{t}_h\|^2_{[L^2(T)]^2} + \|f+\nabla\cdot\boldsymbol{\sigma}_h\|^2_{L^2(T)} + \|\boldsymbol{\sigma}_h - k(\cdot,|\mathbf{t}_h|)\mathbf{t}_h\|^2_{[L^2(T)]^2} \\ + \sum_{e\in E(T)\cap E(\Gamma)} h_e\left(\|g-u_h\|^2_{L^2(e)} + \|\frac{d}{ds}(g-u_h)\|^2_{L^2(e)}\right) \tag{5}$$

where $E(T)$ denotes the set of edges of T and $E(\Gamma)$ the set of edges in the boundary.

3. Results and Discussion

The adaptive algorithm was tested on a singular example over the unit square with a singularity in its upper right corner. The exact solution is $u(x,y) = (2.1-x-y)^{-1/3}$. In Figure 1b, we can see that the algorithm refines the mesh near the singularity. Figure 1c shows a graph that compares, for both uniform and adaptive refinements, the exact error and the defined indicator vs. degrees of freedom (dof). With the adaptive algorithm the total error is lower, and a good agreement between error and indicator is seen. In fact, the efficiency index, which is the quotient between indicator and total error, tends to one as the mesh size decreases.

These simulations were carried out in FreeFEM++, following the mesh refinement algorithm proposed in [3], and the nonlinear system was solved with Newton's method with a tolerance of 10^{-9}, needing four iterations.

Author Contributions: Theoretical analysis, M.G.; numerical experiments, H.V. All authors have read and agreed to the published version of the manuscript.

Funding: The authors acknowledge the support of CITIC (FEDER Program and grant ED431G 2019/01). The research of M.G. is partially supported by the Spanish Ministerio de Economía y Competitividad Grant MTM2016-76497-R, and by Xunta de Galicia Grant GRC ED431C 2018-033. The research of H.V. is partially supported by Xunta de Galicia grant ED481A-2019/413170 and Ministerio de Educación grant FPU18/06125.

Conflicts of Interest: The authors declare no conflict of interest.

References

1. Gatica, G.N.; Heuer, N.; Meddahi, S. On the numerical analysis of nonlinear twofold saddle point problems. *IMA J. Numer. Anal.* **2003**, *23*, 301–330.
2. González, M.; Varela, H. A posteriori error analysis of an augmented dual-mixed method of a nonlinear Dirichlet problem. In preparation (2020).
3. Hecht, F. New development in freefem++. *J. Numer. Math.* **2012**, *20*, 251–265.

Proceedings

European and American Options Valuation by Unsupervised Learning with Artificial Neural Networks †

Beatriz Salvador [1,*], Cornelis W. Oosterlee [2,3] and Remco van der Meer [2,3]

1 Universidade da Coruña, 15001 A Coruña, Spain
2 CWI—Centrum Wiskunde & Informatica, 1098 Amsterdam, The Netherlands; C.W.Oosterlee@cwi.nl (C.W.O.); remcovdmeer0@gmail.com (R.v.d.M.)
3 DIAM, Delft University of Technology, 2628 Delft, The Netherlands
* Correspondence: beatriz.salvador@udc.es

† Presented at the 3rd XoveTIC Conference, A Coruña, Spain, 8–9 October 2020.

Published: 19 August 2020

Abstract: Artificial neural networks (ANNs) have recently also been applied to solve partial differential equations (PDEs). In this work, the classical problem of pricing European and American financial options, based on the corresponding PDE formulations, is studied. Instead of using numerical techniques based on finite element or difference methods, we address the problem using ANNs in the context of unsupervised learning. As a result, the ANN learns the option values for all possible underlying stock values at future time points, based on the minimization of a suitable loss function. For the European option, we solve the linear Black–Scholes equation, whereas for the American option, we solve the linear complementarity problem formulation.

Keywords: (non)linear PDEs; Black–Scholes model; artificial neural network; loss function; multi-asset options

1. Introduction

The interest in machine learning techniques, due to the remarkable successes in different application areas, is growing exponentially. ANNs are learning systems based on a collection of artificial neurons that constitute a connected network. Such systems "learn" to perform tasks, generally without being programmed with task-specific rules. Many different financial problems have also been addressed with machine learning. The financial application on which we focus is the valuation of financial derivatives with PDEs. Generally, we can distinguish between supervised and unsupervised machine learning techniques. The goal of the current work is to solve the financial PDEs by applying unsupervised machine learning techniques. In such a case, only the inputs of the network are known, and based on a suitable loss function that needs to be minimized, the ANN should "converge" to the solution of the PDE problem. We will price European and American options modeled by the Black–Scholes PDE and look for solutions for all future time points and stock values. Thus, linear and nonlinear partial differential equations need to be solved.

2. Artificial Neural Networks Solving PDEs

We introduce the methodology following [1] to solve linear and nonlinear time-dependent PDEs by ANNs. Then, we write a general PDE problem as follows:

$$\begin{aligned}
\mathcal{N}_I(v(t,x)) &= 0, \quad x \in \widetilde{\Omega},\ t \in [0,T], \\
\mathcal{N}_B(v(t,x)) &= 0 \quad \text{on } \partial\widetilde{\Omega}, \\
\mathcal{N}_0(v(t^*,x)) &= 0 \quad x \in \widetilde{\Omega} \text{ and } t^* = 0 \text{ or } t^* = T,
\end{aligned} \tag{1}$$

where $v(t,x)$ denotes the solution of the PDE, $\mathcal{N}_I(\cdot)$ is a linear or nonlinear time-dependent differential operator, $\mathcal{N}_B(\cdot)$ is a boundary operator, $\mathcal{N}_0(\cdot)$ is an initial or final time operator, $\widetilde{\Omega}$ is a subset of $\mathbb{R}^D$, and $\partial\widetilde{\Omega}$ denotes the boundary on the domain $\widetilde{\Omega}$. The goal is to obtain $\hat{v}(t,x)$ by minimizing a suitable loss function $L(v)$ over the space of k-times differentiable functions, where k depends on the order of the derivatives in the PDE, i.e.,

$$arg \min_{v \in C^k} L(v) = \hat{v}$$

where we denote by $\hat{v}(t,x)$ the true solution of the PDE. A general expression for the loss function, defined in terms of the L^p norm, including a weighting, is defined as follows [1,2]:

$$L(v) = \lambda \int_{\Omega} \mid \mathcal{N}_I(v(t,x)) \mid^p d\Omega + (1-\lambda) \int_{\partial\Omega} (\mid \mathcal{N}_B(v(t,x)) \mid^p + \mid \mathcal{N}_0(v(t,x)) \mid^p)\, d\gamma, \tag{2}$$

where $\Omega = \widetilde{\Omega} \times [0,T]$ and $\partial\Omega$ the boundary of Ω. Financial options with early-exercise features give rise to free boundary PDE problems. We will focus on the reformulation of the free boundary problem as a LCP. The generic LCP formulation reads:

$$\begin{aligned}
\max(\mathcal{N}_0(v(t,x)), \mathcal{N}_I(v(t,x))) &= 0, \quad x \in \widetilde{\Omega},\ t \in [0,T], \\
\mathcal{N}_B(v(t,x)) &= 0, \quad \text{on } \partial\widetilde{\Omega}, \\
\mathcal{N}_0(v(t^*,x)) &= 0, \quad x \in \widetilde{\Omega} \text{ and } t^* = 0 \text{ or } t^* = T.
\end{aligned}$$

Our expression for the loss function, to solve the linear complementarity problem, is as follows:

$$L(v) = \lambda \int_{\Omega} \mid \max(\mathcal{N}_0(t,x,v), \mathcal{N}_I(t,x,v)) \mid^p d\Omega + (1-\lambda) \int_{\partial\Omega} (\mid \mathcal{N}_B(t,x,v) \mid^p + \mid \mathcal{N}_0(t,x,v) \mid^p)\, d\gamma. \tag{3}$$

3. Financial Derivatives Pricing PDEs

In this section, the option pricing partial differential equation problems are presented. We briefly introduce the European asset model, the American and multi-asset options model being easily extended.

The underlying asset S_t is assumed to pay a constant dividend yield δ, and follows the geometric Brownian motion:

$$dS_t = (\mu - \delta) S_t dt + \sigma S_t dW_t^P \tag{4}$$

where W_t^P is a Brownian motion. Assuming there are no arbitrage opportunities, the European option value follows from the Black–Scholes equation:

$$\begin{cases} \mathcal{L}(v) = \partial_t v + \mathcal{A} v - rv = 0 & S \in \widetilde{\Omega}\ t \in [0,T) \\ v(T,S) = H(S) \end{cases} \tag{5}$$

where operator $\mathcal{A}$ is the classical Black–Scholes operator, and function H denotes the option's payoff. Then, the loss function is defined as:

$$L(v) = \lambda \int_{\Omega} \mid \mathcal{L}(v(t,x)) \mid^p d\Omega \\ + (1-\lambda) \int_{\partial\Omega} (\mid v(t,x) - G(t,x) \mid^p + \mid v(t,x) - H(x) \mid^p) \, d\gamma, \quad (6)$$

where functions G and H denote the values of the spatial boundary conditions and final condition, respectively. The integral terms in the loss function are approximated by Monte Carlo techniques.

4. ANN Option Pricing Results

We start with a European put option, with the following parameters' values: $\sigma = 0.25$, $r = 0.04$, $T = 1$, $K = 15$, $S_\infty = 4K$, $\delta = 0.0$. In Figure 1, the ANN-based, trained, and the analytical solution are plotted for two time instances. The relative error, with $\lambda = 0.5$, is equal to 2.23×10^{-4}.

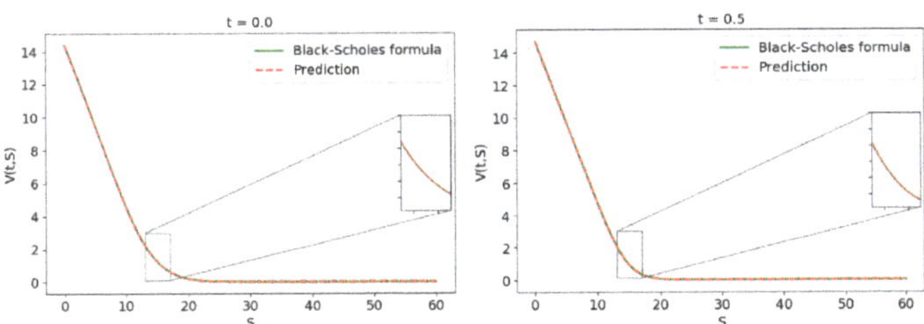

Figure 1. European put option for different times instances, $t = 0, t = 0.5$, with $\lambda = 0.5$.

Finally, in order to show the accuracy of the method applied to train the ANN to price American options depending on two asset prices, the relative error is presented in Table 1.

Table 1. Error for different multi-asset American options.

	Error
Max-call	1.73×10^{-3}
Spread	2.45×10^{-3}
Arithmetic average put	6.42×10^{-3}

Author Contributions: Investigation and writing—original draft preparation: B.S., Supervision: C.W.O., validation: R.v.d.M. All authors have read and agreed to the published version of the manuscript.

Funding: This research was funded by ERCIM fellowship.

Conflicts of Interest: The authors declare no conflict of interest.

References

1. van der Meer, R.; Oosterlee, C.; Borovykh, A.T. Optimally weighted loss function for solving PDEs with neural Networks. *arXiv* **2020**, arXiv:2002.06269.
2. Raissi, M.; Karniadakis, G.E.; Perdikris, P. Physics informed deep learning (part I): Data-driven solutions of nonlinear partial differential equations. *arXiv* **2017**, arXiv:1711.10561v1.

Proceedings

Digital Image Quality Prediction System †

Nereida Rodriguez-Fernandez [1,2,*], Iria Santos [1,2], Alvaro Torrente-Patiño [3] and Adrian Carballal [1,3]

1 CITIC-Research Center of Information and Communication Technologies, University of A Coruña, 15071 A Coruña, Spain; iria.santos@udc.es (I.S.); adrian.carballal@udc.es (A.C.)
2 Department of Computer Science and Information Technologies, Faculty of Communication Science, University of A Coruña, Campus Elviña s/n, 15071 A Coruña, Spain
3 Department of Computer Science and Information Technologies, Faculty of Computer Science, University of A Coruña, Campus Elviña s/n, 15071 A Coruña, Spain; alvaro.torrente@udc.es
* Correspondence: nereida.rodriguezf@udc.es

† Presented at the 3rd XoveTIC Conference, A Coruña, Spain, 8–9 October 2020.

Published: 19 August 2020

Abstract: "A picture is worth a thousand words." Based on this well-known adage, we can say that images are important in our society, and increasingly so. Currently, the Internet is the main channel of socialization and marketing, where we seek to communicate in the most efficient way possible. People receive a large amount of information daily and that is where the need to attract attention with quality content and good presentation arises. Social networks, for example, are becoming more visual every day. Only on Facebook can you see that the success of a publication increases up to 180% if it is accompanied by an image. That is why it is not surprising that platforms such as Pinterest and Instagram have grown so much, and have positioned themselves thanks to their power to communicate with images. In a world where more and more relationships and transactions are made through computer applications, many decisions are made based on the quality, aesthetic value or impact of digital images. In the present work, a quality prediction system for digital images was developed, trained from the quality perception of a group of humans.

Keywords: machine learning; genetic algorithm; quality; image; prediction; dataset

1. Introduction

In recent years, significant efforts were applied to the development of successful models and algorithms that can automatically and accurately predict the perceptual quality of two-dimensional (2D) and three-dimensional (3D) digital images and videos. This estimate comes from studies with at least a century of experience, or more if we take into account those developed by Platon and Aristotle, usually from Humanities departments: Psychology, Sociology, Philosophy, Fine Arts, etc. [1]. Different research groups sought to create computer systems capable of learning the aesthetic and quality perception of a group of humans as part of a generative system for uses such as the selection and arrangement of images within a set, even though it is complex to translate this into computer problems. Visual quality refers to the quantification of the perceptual degradation of a visual stimulus due to the presence or absence of distortions. Most of the applications that were developed were designed to treat synthetically distorted images [2]. In this case, unlike other image quality assessment algorithms that use synthetically distorted images [3,4], it was decided to use images with absence of distortion [5,6]. Despite the fact that the data collected contained quality and aesthetic results, on this occasion only the quality data were used as they constituted more objective results [7].

2. Materials and Methods

After analyzing the degree of generalization of some datasets used in automatic image prediction, it was concluded that it was not enough to consider them as a reference in the training of automatic image prediction and classification systems. Taking this into account, a new set of images from the web portal DPChallenge.com was developed in search of greater statistical consistency [7].

The proposed dataset was built following the steps outlined in previous works [8]: obtaining the images on the web portal, filtering those images, organizing them according to their evaluation on the portal and selecting sets with an equal number of images. Subsequently, the quality of the images was evaluated by a group of humans through the Amazon Mechanical Turk platform. This group of humans was made up of 525 inhabitants of the USA (39% men and 61% women), aged between 18 and 70. A representation of the images from this dataset is shown in Figure 1.

Figure 1. Images of different scoring ranges belonging to the dataset used in this work, evaluated by humans according to their quality. (**a**) Image with an average score of 2.7 out of a maximum of 10. (**b**) Image with an average score of 5.7 out of a maximum of 10. (**c**) Image with an average score of 9.28 out of a maximum of 10.

With this data, a system was created to predict the quality of digital images with the search engine Correlation by Genetic Search (CGS) [9,10].

3. Results

The results obtained during the experimental phase correspond to 50 runs of a 5-fold cross-validation with a training model where 80% of the set is dedicated to training and the remaining 20% to testing. As input data, 1024 features of VGG19 were used. The average number of features used in the 50 runs is 114, which has also reached an average Pearson correlation of 0.77 and an average error of 0.15. Figure 2 shows the distribution of features, Pearson correlation and error of the 50 runs. The absence of a large number of outliers stands out, which provides consistency and validity to the data obtained. In the three cases, the data that is recognized as outlier belongs to the same run. In the case of the error, its greater variability can be observed, with a maximum error of 0.16 and a minimum error of 0.09. In the case of the features and the Pearson correlation, a much more uniform and concentrated representation is observed, with a very small variability that leads to deduce that the model proposes coherent results in the 50 runs.

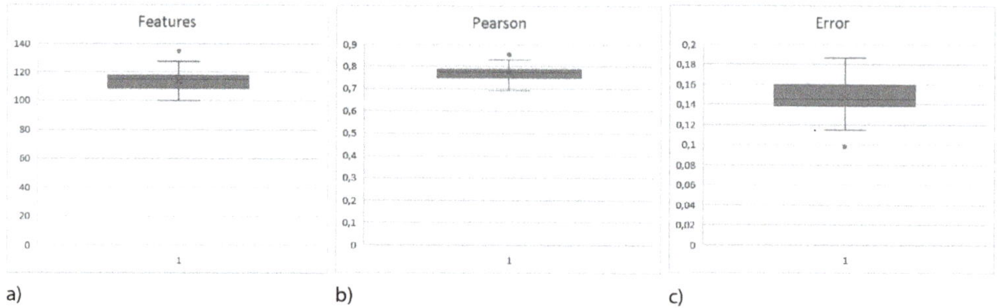

Figure 2. Results obtained in the experiments carried out with CGS. (**a**) Features used in each of the 50 runs. (**b**) Pearson correlation of the validation set in each run. (**c**) Average error obtained in each run.

4. Conclusions

This paper focuses on the creation of a digital image quality prediction system from a set of human-evaluated images. The task was tested with a hybrid method for the creation of multiple regression models based on the maximization of the correlation, the CGS method. Thus, an average Pearson correlation of 0.77 in 50 runs with 5-fold cross-validation was achieved, with a consistent distribution and low variability, which provides better results than other state-of-the-art works such as Nadal et al. [11] or Marin and Leder [12].

Acknowledgments: CITIC, as a Research Centre of the Galician University System, is financed by the Regional Ministry of Education, University and Vocational Training of the Xunta de Galicia through the European Regional Development Fund (ERDF) with 80%, Operational Programme ERDF Galicia 2014-2020 and the remaining 20% by the General Secretariat of Universities (Ref. ED431G 2019/01). This work has also been supported by the General Directorate of Culture, Education and University Management of Xunta de Galicia (Ref. ED431D 201716), and Competitive Reference Groups (Ref. ED431C 201849)

References

1. Maitre, H. A Review of Image Quality Assessment with application to Computational Photography. In Proceedings of the Ninth International Symposium on Multispectral Image Processing and Pattern Recognition, Enshi, China, 31 October–1 November 2015.
2. Celona, L. Learning Quality, Aesthetics, and Facial Attributes for Image Annotation. Tesi di Dottorato, Universitàdegli Studi di Milano-Bicocca, Milan, Italy, 2018.
3. Li, J.; Yan, J.; Deng, D.; Shi, W.; Deng, S. No-reference image quality assessment based on hybrid model. *Signal Image Video Process.* **2016**, *11*, doi:10.1007/s11760-016-1048-5.
4. Xu, J.; Ye, P.; Li, Q.; Du, H.; Liu, Y.; Doermann, D. Blind image quality assessment based on high order statistics aggregation. *IEEE Trans. Image Process.* **2016**, *25*, 4444–4457.
5. Ghadiyaram, D.; Bovik, A. Automatic quality prediction of authentically distorted pictures. *Proc. SPIE* **2015**, doi:10.1117/2.1201501.005759.
6. Ghadiyaram, P.D. Perceptual Quality Assessment of Real-World Images and Videos. Ph.D. Thesis, The University of Texas, Austin, TX, USA, 2017.
7. Rodriguez-Fernandez, N.; Santos, I.; Torrente, A. Dataset for the Aesthetic Value Automatic Prediction. *Proceedings* **2019**, *21*, 31.
8. Carballal, A.; Fernandez-Lozano, C.; Rodriguez-Fernandez, N.; Castro, L.; Santos, A. Avoiding the inherent limitations in datsets used for measuring aesthetics whtn using a machine learning approach. *Complexity* **2019**, 1–12, doi:10.1155/2019/4659809.
9. Carballal, A.; Fernandez-Lozano, C.; Heras, J.; Romero, J. Transfer learning for predicting aesthetics through a novel hyprid machine learning method. *Neural Comput. Appl.* **2019**, *32*, 5889–5900.

10. Pazos Perez, R.I.; Carballal, A.; Rabuñal, J.R.; Mures, O.A.; García-Vidaurrázaga, M.D. Predicting vertical urban growth using genetic evolutionary algorithms in Tokyo's Minato ward. *J. Urban Plan. Dev.* **2018**, *144*, 04017024.
11. Nadal, M.; Munar, E.; Marty, G.; Cela-Conde, C. Visualcomplexity and beauty appreciation: Explaining the divergenceof results. *Empir. Stud. Arts* **2010**, *28*, 173–191.
12. Marin, M.; Leder, H. Examining complexity acrossdomains: Relating subjective and objective measures of affectiveenvironmental scenes, paintings and music. *PLoS ONE* **2013**, *8*, e72412.

Proceedings

Radiology Seminars with Guest Professors in the Virtual Environment Second Life®: Perception of Learners and Teachers †

Juan M. Alonso-Martinez [1,*], Shaghayegh Ravaei [1], Teodoro Rudolphi-Solero [2] and Francisco Sendra-Portero [1,*]

[1] Department of Radiology and Physical Medicine, School of Medicine, University of Málaga, 29016 Málaga, Spain; shaghayegh.ravaei@gmail.com
[2] Department of Nuclear Medicine, University Hospital Virgen de las Nieves, 18014 Granada, Spain; teorudsol@gmail.com
* Correspondence: martin17alonso@gmail.com (J.M.A.-M.); sendra@uma.es (F.S.-P.)
† Presented at the 3rd XoveTIC Conference, A Coruña, Spain, 8–9 October 2020.

Published: 19 August 2020

Abstract: Nine professors of radiology from six different cities were invited to give a 1-hour seminar in the virtual world Second Life® to 154 third-year medical students from the University of Málaga. Students and teachers performed a questionnaire about the cognitive load that implies receiving/teaching seminars inside Second Life® and several characteristics involving the experience. This experience was considered remarkably enriching by teachers and learners and opens new interesting pathways for educational contact between students and teachers from different universities, with the advantages of reducing costs and travel time.

Keywords: online learning; virtual worlds; training of trainers; medical students; undergraduate education; radiology

---

1. Introduction

Second Life® (Linden Research SL, San Francisco, CA, USA) is one of the most complex and creative virtual world platforms, with the potential to engage students in learning processes that enhance creative collaboration [1]. Different learning activities, organized by the Department of Radiology and Physical Medicine of the University of Malaga, have been carried out within Second Life® since 2011 and have mainly focused on the perception of students and the impact on their learning [2–5], but there have been no approaches focused on the perspective of teachers. The objective of this study was to explore the teaching of professional Radiologists with none or little previous experience in Second Life® to third-year medical students, during the development of the core course, Radiology, as well as to evaluate the perceptions of students and professors.

2. Materials and Methods

Several professors were invited to participate in one-hour seminars organised between 12 March and 27 May, with the only prerequisite being that they did not work for the Radiology Department of the University of Málaga. As a result, 9 medical specialists in Radiology or Nuclear Medicine from 6 different cities (Badajoz, Cádiz, Córdoba, Granada, Madrid and Málaga) did participate. Only one of them had some previous experience teaching in Second Life®. The remaining eight received prior training to correctly use the main basic functions of the platform (moving, speaking, audio functions, using the camera and the avatar view and controlling the slide-show presentation). The topics, arranged in chronological order, were as follows:

1. Introduction to nuclear medicine: a millennial survival guide
2. Atelectasis: basic aspects for students
3. Advanced CNS image: from form to function
4. Abdominal CT and: what we see in each technique
5. Solitary bone lesion
6. Therapeutic procedures and diagnostic techniques in musculoskeletal radiology
7. A talk of chest and abdominal X-rays: where to look.
8. Radiology of the musculoskeletal system. Pasapalabra?
9. Basic concepts and radiological aspects of solitary bone lesions

Both students and teachers were invited to perform an evaluation questionnaire about the cognitive load (mental effort) required in this activity as well as diverse characteristics involving the experience, the 3D platform itself and the given or received seminar. Cognitive load was measured on a 9-point scale [6], answering the question "How much mental effort does it cost you to function in Second Life, considering: (1) very, very low mental effort; (2) very low mental effort; (3) low mental effort; (4) somewhat low mental effort; (5) neither much nor little mental effort; (6) somewhat high mental effort; (7) high mental effort; (8) very high mental effort; and (9) very, very high mental effort. The remaining items of the questionnaire were scored from 0 to 10.

(a)

(b)

Figure 1. Screenshots during the radiology seminars: (**a**) Aerial view of the island where the seminars were held, specifically on the floating platform over the central area of the island. (**b**) Scene during one of the seminars given, in which students can be seen sitting in the open-air auditorium in front of the slide show screen while the teacher makes his presentation.

Table 1. Results of the questionnaire about the experience [1].

Assessed Item	Students	Teachers	p [2]
Cognitive load [3]	3.6 ± 1.8	4.9 ± 1.6	0.049
Connectivity to Second Life [4]	8.1 ± 1.7	9.3 ± 1.2	0.056
The environment of the island [4]	9.3 ± 1.1	9.5 ± 0.9	0.555
Audio quality [4]	7.9 ± 2.0	8.9 ± 1.6	0.164
The experience receiving/teaching this seminar [4]	8.6 ± 1.4	9.6 ± 0.7	0.049
The teacher's assessment/self-assessment [4]	9.2 ± 1.3	7.6 ± 1.8	0.001
The interest of content for third-year students [4]	9.0 ± 1.3	8.4 ± 1.8	0.189
Assessment/self-assessment of oral exposure [4]	9.0 ± 1.3	7.3 ± 1.7	<0.001
The quality of the graphic presentation (slideshow) [4]	9.1 ± 1.3	8.3 ± 1.0	0.083

[1.] Results are mean ± standard deviation of all given values from both populations students and teachers. [2.] p values from unpaired Student-t test, $p < 0.05$ was considered significative difference. [3] The cognitive load was requested on a 9-point Likert scale [6]. [4.] The remaining items were scored from 0 to 10.

3. Results

One hundred and fifty-four students attended the seminars: 152 new students (86.4%) and 2 repeating students (4.5%). The number of attendees per seminar ranged between 21 and 39 (mean 28.8 ± 6.2). Seventy-two students attended one seminar, 63 attended two while the remaining attended at least three. Both professors and students evaluated the experience as remarkably positive and a high percentage reported they would be willing to repeat a similar approach.

4. Discussion

Although Second Life® was created in 2003, it is still little known within the medical university educational area. Second Life® has numerous advantages, two of them being the realistic approach inside the platform, and the ubiquitous perception acquired through remote access. The cognitive load was significantly lower for students than for teachers, which may be related to the intergenerational difference in technological literacy and the adaptability to new forms of communication.

This experience was remarkably enriching for both professors and students. From the students' perspective, the selected seminars play an important complementary role on their Radiology formation (e.g., expanding their radiology conceptual framework and fixing essential concepts exposed throughout the main core course). In addition, students acquired a sense of professional growth approaching their first medical congress "outside the classic medical school environment". From the perspective of the invited professors, this project has been a training of trainers experience, allowing them to learn about a new online infrastructure with interesting capabilities of synchronous teaching and compare it (time, accessibility, ubiquity, synchrony, facilities, sound, 3D image, etc.) with the traditional method and other 2D online platforms. This experience opens new interesting and enriching pathways for educational contact between students and teachers from different universities, with the advantages of reducing costs and travel time.

Author Contributions: Conceptualization, F.S.-P.; methodology, J.M.A.-M., S.R., T.R-S. and F.S.-P.; software, F.S.-P.; validation, J.M.A.-M., S.R., T.R-S. and F.S-P; formal analysis, J.M.A.-M., S.R., T.R-S. and F.S.-P.; investigation, J.M.A.-M., S.R., T.R-S. and F.S.-P.; resources, F.S.-P.; data curation, J.M.A-M and F.S.-P.; writing—original draft preparation, J.M.A-M and F.S.-P.; writing—review and editing, J.M.A.-M., S.R., T.R-S. and F.S.-P.; visualization, J.M.A.-M., S.R., T.R-S. and F.S.-P.; supervision, F.S.-P.; project administration, F.S.-P.; funding acquisition, F.S.-P. All authors have read and agreed to the published version of the manuscript.

Funding: The Innovative Education Project #PIE19-217 of the University of Málaga partially supported this study. The maintenance cost of the Medical Master Island during this project was supported by the Andalusian Society of Radiology (Asociación de Radiólogos del Sur), a subsidiary of the Spanish Society of Medical Radiology (SERAM).

Acknowledgments: We want to express our appreciation to the teachers and students who have agreed to participate in this teaching experience.

Conflicts of Interest: The authors declare no conflict of interest.

References

1. Olteanu, R.L.; Bîzoi, M.; Gorghiu, G.; Suduc, A.M. Working in the Second Life environment: a way for enhancing students' collaboration. *Procedia Soc. Behav. Sci.* **2014**, *141*, 1089–1094
2. Sendra-Portero, F.; Lorenzo-Alvarez, R.; Pavia-Molina, J. Teaching radiology in the "Second life" virtual world. *Diagn. Imag. Eur.* **2018**, *34*, 43–45.
3. Lorenzo-Alvarez, R.; Pavia-Molina, J.; Sendra-Portero, F. Exploring the potential of undergraduate radiology education in the virtual world Second Life with first-cycle and second-cycle medical students. *Acad. Radiol.* **2018**, *25*, 1087–1096.
4. Lorenzo-Alvarez, R.; Ruiz-Gomez, M.J.; Sendra-Portero, F. Medical students' and family physicians' attitudes and perceptions toward radiology learning in the virtual world Second Life. *AJR Am. J. Roentgenol.* **2019**, *212*, 1295–1302.
5. Lorenzo-Alvarez, R.; Rudolphi-Solero, T.; Ruiz-Gomez, M.J.; Sendra-Portero, F. Medical student education for abdominal radiographs in a 3D virtual classroom versus traditional classroom: A randomized controlled trial. *AJR Am. J. Roentgenol.* **2019**, *213*, 644–650.

6. Paas, F.; van Merriënboer, J.J.G. Instructional control of cognitive load in the training of complex cognitive tasks. *Educ. Psychol. Rev.* **1994**, *6*, 51–71.

Proceedings

Preliminary Analysis of a Virtual Inter-University Game to Learn Radiology within the Second Life® Environment †

Alberto-Jimenez-Zayas *, Shaghayegh Ravaei, Juan M. Alonso-Martinez and Francisco Sendra-Portero *

Department of Radiology and Physical Medicine, School of Medicine, University of Málaga, 29071 Málaga, Spain; shaghayegh.ravaei@gmail.com (S.R.); martin17alonso@gmail.com (J.M.A.-M.)

* Correspondence: albertojzayas@hotmail.es (A.J.-Z.); sendra@uma.es (F.S.-P.)

† Presented at the 3rd XoveTIC Conference, A Coruña, Spain, 8–9 October 2020.

Published: 19 August 2020

Abstract: A competition-based game, named League of Rays (LOR), designed to learn radiology within the multi-user virtual environment Second Life was adapted for the participation of teams of four students. The game ran from 20 February to 1 April 2020. Forty-one teams from 16 universities initially signed up and 28 teams from 14 universities finished the game. Participants found this activity fun, enjoyable and useful for their training. Some interesting proposals to be included in future editions of the game and interesting comments on the meaning of developing half of the game during the confinement caused by the Covid-19 pandemic were provided from participants.

Keywords: online learning; virtual worlds; gamification; game-based learning; medical students; undergraduate education; radiology

---

1. Introduction

In 2015, a virtual game called League of Rays (LOR) was designed to learn radiology in the multi-user immersive environment of Second Life® [1]. After several editions of individual participation, it was considered to replicate the game with students from different medical schools, taking advantage of the remote and free use of Second Life®, adding a "sense of belonging" and social comparison to the competition [2]. In 2019, the rules for participating in teams of four students were developed and in 2020 an edition with interuniversity teams was carried out, with the only requirement that participants were coursing a course on radiology this year. The goal of this study is to carry out a preliminary analysis of this last edition of LOR.

2. Materials and Methods

The game was organized in six weeks, the first three were dedicated to radiological anatomy and the latter three to radiological semiology. Participants had to register in Second Life, make an avatar and interact with the environment of the island, performing the actions required by the game (Figure 1). During the first four days of each week, participants had to view educational content presented in sets of three panels, as self-guided slide shows. On the last three days, the participants had to solve individual assessments (multiple choice test) and a team assessment related to the theme of the week. The score acquired for each team determined their position in the classification of participants. If one of the teams stopped participating for two consecutive weeks, it was disqualified. The organization of the game was in constant contact with participants by email. After the game, participants were asked to complete a questionnaire evaluating the project.

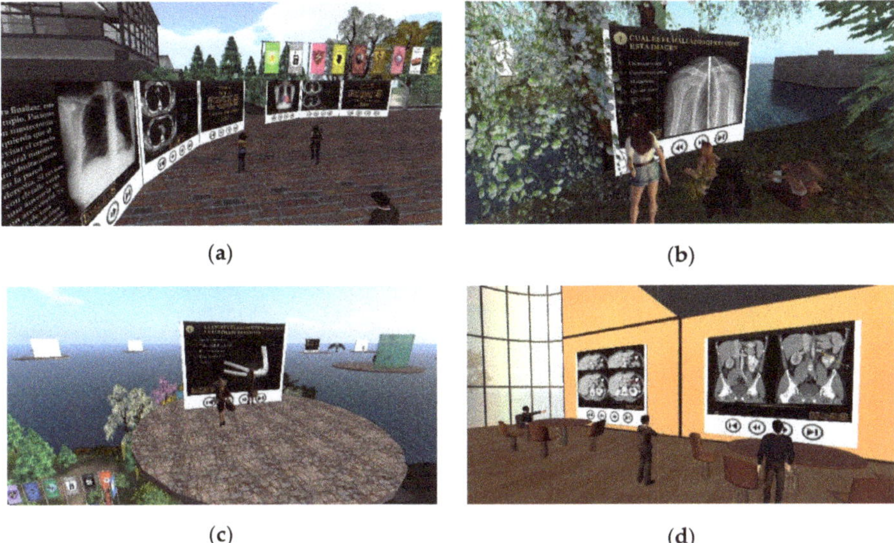

Figure 1. Several screenshots of the 2020 edition of the game League of Rays: (**a**) students watching educational content presented in sets of three panels; (**b**) a group of participants in front of the panel with the multiple-choice test corresponding to one of them. (**c**) Students doing an individual assessment on floating platforms in the sky; (**d**) several participants reviewing the assessment tasks by team.

3. Results

The game ran from 20 February to 1 April 2020, so the second half of it took place during confinement due to the Covid-19 pandemic. Forty-one teams from 16 universities initially signed up and 28 teams from 14 universities finished the game. The questionnaire was completed by 93 participants (83.0% of those who completed the game). The results are summarized in Figure 2. The students rated with average scores greater than or equal to 8.2 points out of 10, highlighting the organization of the project, the teacher, the usefulness of their information and interaction with peers (average scores greater than 9).

Figure 2. Bar graph depicting the mean rating of various aspects of the League of Rays game scored on a scale of 0 to 10. Error bars represent standard deviation. N or P means normal or pathological.

4. Discussion

The Second Life® virtual platform can be accessed free of charge and the learning opportunities it offers are almost endless, allowing communication, meeting and contact between different professionals in a synchronic and asynchronous way. LOR is intended as an asynchronous game-based learning approach. Since its first edition in 2015, it has proven to be a well-accepted game-based learning experience for students, which allows learning basic aspects of radiological anatomy and semiology in a very motivating way, providing a complementary tool to engage students in radiology learning [1]. The last 2020 edition also encourages teamwork and competitive pressure as a stimulus to establish radiology concepts and skills. The call can be considered very satisfactory, since 112 students from 14 different universities finally completed the activity.

Students increase their engagement and participation in their education through game-based learning and virtual worlds [3,4]. Current medical students, belonging to Generation Z, may find game-based learning activities, such as LOR, particularly attractive for learning because they are active problem solvers, independent learners and appreciate healthy competition [5]. This activity, developed as a multi-user online game, has been fun, enjoyable and useful for their training. Participants provided interesting proposals to be included in future editions of the game and interesting comments on the meaning of developing half of the game during the confinement by the Covid-19 pandemic.

Supplementary Materials: A PDF with information about the LOR 2020 prior to the start of the game is available online at http://www.biznaga.org/LOR2020/LOR2020.pdf. The last qualification of the game is available online at http://www.biznaga.org/LOR2020/LOR2020-37-SEXTA-calcificacion-general-ord.pdf.

Author Contributions: Conceptualization, F.S.-P.; methodology, A.J-Z., S.R., J.M.A.-M. and F.S.-P.; software, F.S.-P.; validation, A.J.-Z. and F.S.-P.; formal analysis, A.J.-Z. and F.S.-P.; investigation, A.J.-Z., S.R., J.M.A.-M. and F.S.-P.; resources, F.S.-P.; data curation, A.J.-Z.; writing—original draft preparation, A.J.-Z.; writing—review and editing, A.J.-Z. and F.S.-P.; visualization, A.J.-Z., S.R., J.M.A.-M and F.S.-P.; supervision, F.S.-P.; project administration, F.S.-P.; funding acquisition, F.S.-P. All authors have read and agreed to the published version of the manuscript.

Funding: The Innovative Education Project #PIE19-217 of the University of Málaga partially supported this study. The maintenance cost of the Medical Master Island during this project was supported by the Andalusian Society of Radiology (Asociación de Radiólogos del Sur), a subsidiary of the Spanish Society of Medical Radiology (SERAM).

Acknowledgments: We want to thank all the students who have participated in this game-based learning experience, as well as their teachers, who have encouraged them to do so.

Conflicts of Interest: The authors declare no conflict of interest.

References

1. Lorenzo-Alvarez, R.; Rudolphi-Solero, T.; Ruiz-Gomez, M.J.; Sendra-Portero, F. Game-based learning in virtual worlds: A multiuser online game for medical undergraduate radiology learning within Second Life. *Anat. Sci. Educ.* **2019**, doi:10.1002/ase.1927
2. Van Nuland, S.E.; Roach, V.A.; Wilson, T.D.; Belliveau, D.J. Head to head: The role of academic competition in undergraduate anatomical education. *Anat. Sci. Educ.* **2015**, *8*, 4–412.
3. Brigham, T.J. An introduction to gamification: Adding game elements for engagement. *Med. Ref. Serv. Q.* **2015**, *34*, 471–480.
4. Gong W. Education and Three-Dimensional Virtual Worlds: A Critical Review and Analysis of Applying Second Life in Higher Education. Master's Thesis, University of British Columbia, Vancouver, BC, Canada, 2018.
5. Eckleberry-Hunt, J.; Lick, D.; Hunt, D. Is Medical Education Ready for Generation Z? *J. Grad. Med. Educ.* **2018**, *10*, 378–381.

Proceedings

Introduction to Second Life® As a Virtual Training Environment: Perception of University Teachers †

Teodoro Rudolphi-Solero [1,*], Alberto Jimenez-Zayas [2] and Francisco Sendra-Portero [2,*]

[1] Department of Nuclear Medicine, University Hospital Virgen de las Nieves, 18014 Granada, Spain
[2] Department of Radiology and Physical Medicine, School of Medicine, University of Málaga, 29071 Málaga, Spain; albertojzayas@hotmail.es
* Correspondence: teorudsol@gmail.com (T.R.-S.); sendra@uma.es (F.S.-P.)
† Presented at the 3rd XoveTIC Conference, A Coruña, Spain, 8–9 October 2020.

Published: 20 August 2020

Abstract: In January 2020, two three-hour workshops on an introduction to Second Life® as an online educational platform were held inside a virtual world. The workshops were dedicated to medical university teachers with the main objective being to let them get to know Second Life® and its formative possibilities. The format of this experience was well received by the participants. Everyone who answered the questionnaire agreed that the environment was useful and interesting and that they would repeat a similar experience again. Ten out of 23 participants (43.5%) declared that they were willing to carry out a teaching activity in Second Life®. This kind of action allows for the promotion of other future in-world actions directed to the training of trainers.

Keywords: online learning; virtual worlds; training of trainers; undergraduate education; radiology

---

1. Introduction

Second Life® is one of the most well-known multiuser virtual worlds for higher education [1,2]. Since 2011, different teaching activities have been developed in the Medical Master Island, a place inside Second Life® dedicated to medical education [3,4] that, by the end of 2019, involved more than 1800 undergraduate and postgraduate students. In parallel to the Second Life® teaching expansion among medical students [5–7], it is also important to promote the knowledge of Second Life® among other medical teachers. The objective of this study was to evaluate the perceptions of medical teachers about Second Life® after an introductory session to the 3D virtual world and its educational possibilities in medicine.

2. Materials and Methods

Two three-hour-length sessions were held in The Medical Master Island on January 23rd and 30th 2020. The sessions consisted of: (1) a basic training in the main functions of the avatar (walking, flying, speaking, listening, chatting, sending note cards and controlling the settings to see the virtual world); (2) a guided visit to the island and its facilities; (3) a conference about the activities held at the Medical Master Island since 2011; (4) a practical presentation about the technical possibilities of Second Life®, useful resources and other previous experiences in this 3D environment (Figure 1). An invitation was sent in December 2019, along with the program of the sessions, through the academic mailing lists of the Medical School of Málaga and the Association of University Professors of Radiology and Physical Medicine (APURF). A PDF with a basic guide to Second Life® was provided to the attendees before the workshop (see Supplementary Materials).

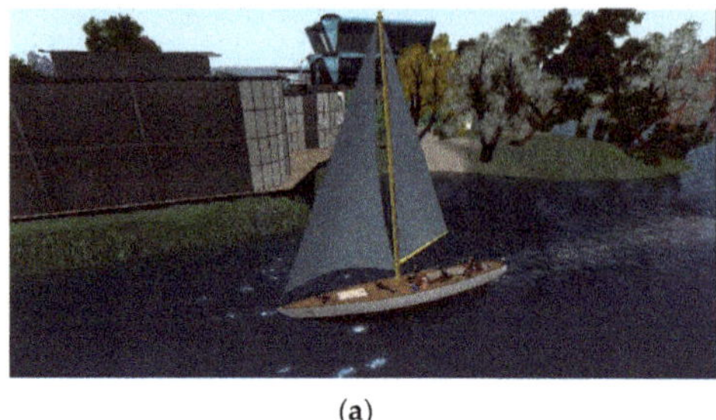

(a)

(b)

Figure 1. Two screenshots of the workshop for medical teachers held in the virtual place "The Medical Master Island": (**a**) A guided visit by boat around the island; (**b**) A session about technical possibilities of Second Life, rendering objects and sculpted figures, and explaining the activity of the former "Isla de la Salud" owned by the Spanish Association of Family and Community Medicine (SAMFYC).

Attendees were asked to send a notecard to the organizer at the beginning of the session, evaluating the cognitive load involved in handling Second Life, using the 9-point cognitive load scale, developed by Paas and van Merriënboer [8] and explaining in brief their first impressions about the experience. After the session, they were asked for a second notecard with four 1- to 5-point Likert scale questions about their perception of Second Life and their willingness to participate in the 3D world as teachers, with a space for "any additional comments" in open format.

3. Results

Twenty-three teachers from seven different universities (Málaga, Córdoba, Granada, Sevilla, Autónoma of Barcelona, Salamanca and La Coruña) participated in the workshops. All were mainly Radiology and Physical Medicine teachers (16), although teachers from other knowledge areas, such as Anatomy (1), Pathological Anatomy (1), Biochemistry (1), Pharmacology (2), Otorhinolaryngology (1) and Pediatrics (1) also participated. Regarding cognitive load, on a scale of 1 to 9, one participant indicated that administering Second Life® implied a very, very high mental effort (9 points), eight indicated somewhat high effort (6 points), three indicated neutral effort (5 points), five indicated low or very low effort (2–3 points) and the remaining six did not answer that question. The answers to the questions about Second Life® in a 1–5-point Likert scale are shown in Table 1.

Table 1. Answers of the teachers attending the workshop on the questions about Second Life.

Questions	Likert Scale Values [1]					
	1	2	3	4	5	NA
Did you find the workshop interesting?	-	-	-	2	12	9
Do you think Second Life® is useful for teaching?	-	-	-	4	10	9
Would you be interested in participating in future meetings for teachers?	-	-	-	2	12	9
Would you be interested in participating as a teacher in Second Life®?	-	-	4	1	9	9

[1] Likert scales values were: 1—strongly disagree, 2—disagree, 3—neutral (nor disagree neither agree), 4—agree, 5—strongly agree. NA: not answered.

4. Discussion

The format of this experience was well received by the participants. Everyone who answered the questionnaire agreed that the environment was useful and interesting and that they would repeat similar experiences again. Ten out of 23 participants (43.5%) declared that they were willing to carry out a teaching activity in Second Life®.

The workshop provided a first contact with Second Life® and its formative possibilities to medical teachers who attended. As it is designed, it is a 3-h activity, so it is easy to include into university teachers' and specialist doctors' busy schedule. Although the situation supervened by the Covid-19 home confinement limited the available time to organize new similar workshops (mainly because of the increase in clinical workload), in the future we intend to spread the educational potential of Second Life® among a greater number of teachers of medicine and other health-related sciences. This kind of action allow promote other future in-world actions directed to training of trainers: (i) fostering collaborative work in the Second Life® virtual world; (ii) familiarizing teachers with the educational possibilities of virtual worlds and the basic technology applicable to them; (iii) sharing teaching experiences carried out in Second Life® to achieve new interdisciplinary collaborative projects; (iv) using Second Life® as a means of meeting, gathering and having virtual debates; (v) promoting the development of multidisciplinary and inter-university collaborative educational projects.

Supplementary Materials: The workshop program brochure is available online at http://www.biznaga.org/Programa-IntroSL1.pdf, and the basic guide to Second Life is available online at http://www.biznaga.org/guia-basica-secondlife.pdf.

Author Contributions: Conceptualization, F.S.-P. and T.R.-S.; methodology, F.S.-P. and T.R.-S.; software, F.S.-P. and T.R.-S.; validation, T.R.-S., A.J.-Z. and F.S.-P.; formal analysis, T.R.-S., A.J.-Z. and F.S.-P.; investigation, T.R.-S., A.J.-Z. and F.S.-P.; resources, F.S.-P.; data curation, T.R.-S.; writing—original draft preparation, T.R.-S.; writing—review and editing, T.R.-S., A.J.-Z. and F.S.-P.; visualization, T.R.-S., A.J.-Z. and F.S.-P.; supervision, F.S.-P.; project administration, F.S.-P.; funding acquisition, F.S.-P. All authors have read and agreed to the published version of the manuscript.

Funding: The Innovative Education Project #PIE19-217 of the University of Málaga partially supported this study. The maintenance cost of the Medical Master Island during this project was supported by the Andalusian Society of Radiology (Asociación de Radiólogos del Sur), a subsidiary of the Spanish Society of Medical Radiology (SERAM).

Acknowledgments: We want to thank José Pavia-Molina and Félix del Ojo for their participation and help in the two days of the Second Life introductory workshop for teachers.

Conflicts of Interest: The authors declare no conflict of interest.

References

1. Potkonjak, V.; Gardner, M.; Callaghan, V.; Mattila, P.; Guetl, C.; Petrović, V.M.; Jovanović, K. Virtual laboratories for education in science, technology, and engineering: A review. *Comput. Educ.* **2016**, *95*, 309–327.
2. Warburton, S. Second life in higher education: Assessing the potential for and the barriers to deploying virtual worlds in learning and teaching. *Br. J. Educ. Technol.* **2009**, *40*, 414–426.
3. Sendra-Portero, F.; Lorenzo-Alvarez, R.; Pavia-Molina, J. Teaching radiology in the "Second life" virtual world. *Diagn. Imag. Eur.* **2018**, *34*, 43–45.
4. Lorenzo Álvarez, R.; Pavia-Molina, J.; Sendra-Portero, F. Possibilities of the three-dimensional virtual environment tridimensional Second Life® for training in radiology. *Radiologia* **2018**, *60*, 273–279.
5. Lorenzo-Alvarez, R.; Pavia-Molina, J.; Sendra-Portero, F. Exploring the potential of undergraduate radiology education in the virtual world Second Life with first-cycle and second-cycle medical students. *Acad. Radiol.* **2018**, *25*, 1087–1096.
6. Lorenzo-Alvarez, R.; Ruiz-Gomez, M.J.; Sendra-Portero, F. Medical students' and family physicians' attitudes and perceptions toward radiology learning in the virtual world Second Life. *AJR Am. J. Roentgenol.* **2019**, *212*, 1295–1302.

7. Lorenzo-Alvarez, R.; Rudolphi-Solero, T.; Ruiz-Gomez, M.J.; Sendra-Portero, F. Medical student education for abdominal radiographs in a 3D virtual classroom versus traditional classroom: A randomized controlled trial. *AJR Am. J. Roentgenol.* **2019**, *213*, 644–650.
8. Paas, F.; van Merriënboer, J.J.G. Instructional control of cognitive load in the training of complex cognitive tasks. *Educ. Psychol. Rev.* **1994**, *6*, 51–71.

Proceedings

Design of a System to Implement Occupational Stress Studies Trough Wearables Devices and Assessment Tests †

Patricia Concheiro-Moscoso [1,*], María del Carmen Miranda-Duro [1], Carlota Fraga [1], Cristina Queirós [2], António José Pereira da Silva Marques [3] and Betania Groba [1]

[1] CITIC, TALIONIS Group, Elviña Campus, Universidade da Coruña (University of A Coruna), 15008 A Coruna, Spain; carmen.miranda@udc.es (M.d.C.M.-D.); fs.carlota.fs@gmail.com (C.F.); b.groba@udc.es (B.G.)

[2] Laboratório de Reabilitação, Faculdade de Psicologia e de Ciências da Educação, Universidade do Porto (University of Porto), 4200-374 O Porto, Portugal; cqueiros@fpce.up.pt

[3] Laboratório de Reabilitação Psicossocial, Escola Superior de Saúde do Porto, Instituto Politécnico do Porto (Polytechnic of Porto), 4200-374 O Porto, Portugal; antoniomarques@sc.ipp.pt

* Correspondence: patricia.concheiro@udc.es

† Presented at the 3rd XoveTIC Conference, A Coruña, Spain, 8–9 October 2020.

Published: 20 August 2020

Abstract: Introduction: Stress at work is a factor that has repercussions on both a personal and health level, as well as on productivity at work. **Objective**: To establish if the wearables are devices capable of determining the level of labor stress of working people in a research center. **Methodology:** This pilot study followed up different variables during 6 months on 11 participants of a research center. In the study, wearables Xiaomi MiB and 3 were used, which recorded and continuously monitored the physical activity and sleep of the participants. On the other hand, different specific evaluation tests were used to measure work stress, quality of life and sleep quality. **Results**: The data obtained from the tests and the wearables show that men feel slightly more stressed and sleep worse than women; however, men spend more time sitting and walking than women. **Conclusions:** It is considered important to replicate the study in larger and more heterogeneous cohorts.

Keywords: wearables; work stress; quality of life

---

1. Introduction

Work stress is a phenomenon that has been gaining in importance in the recent years. It is defined as "a harmful reaction that people have to cope with the pressures and undue demands placed on them at work" [1]. It has a high impact and repercussions on personal and health levels, as well as on productivity at work [2]. When the occupational stress situation maintenances and increases gradually over time, it could be Burnout [3]. This syndrome is defined as "a prolonged response to chronic emotional and interpersonal stressors at work, and is integrated by the dimensions of burnout, cynicism and inefficiency" [3]. The Burnout has a negative influence in the employment and employees' psychological welfare [4].

Some researches on the stress detection reported to the need for improvements the measurement of daily stressors and in the design of studies with the aim of knowing the influence that the different stress process have in the quality of life [5]. So, in this direction, wearable devices as phones, activity trackers, or sensors could become the main tools for the continuous and real-time monitoring of different parameters in order to contribute to health processes and make a relevant contribution to research [6]. In addition, these technological devices can measure physical activity and sleep quality to relate to possible patterns of behavior when exposed to significant stress situations. Some studies, such as Liao et al. (2005), Mozos et al. (2016) and Han et al. (2017), monitored stress in real time through combination of sensors systems to detect stressful situations [7].

This study provides to a new form to capture stress signals by using a smart wristband. Therefore, the aim of this research was to stablish if these wearables are devices capable of determining the level of occupational stress of workers in a research center in Galicia.

2. Methods

2.1. Study Design

This pilot study was conducted in a research center during 6 months. It included 11 participants based on the inclusion criteria. The inclusion of participants, who were working in an administrative management department from a research center, is highlighted. Prior to participation, informed consent was obtained from all research volunteers. The study protocol was approved by the Autonomic Research Ethics of A Coruña-Ferrol (2019/249).

2.2. Measures

2.2.1. Assessment Tools

All tools were self-administrated, coded and managed through Consortium Research Electronic Data Capture (REDCAP) software. The tools were designed, configured and assigned to each of participants via REDCAP. Moreover, the researchers followed up on questionnaires that were filled in by the participants during the study.

The sociodemographic questionnaire was self-administrated at the start of the study. Perceived Stress Scale v10, Pittsburgh Sleep Quality Index scale, EuroQol-5D-5L scale were covered at the start and the end of the research.

Weekly questionnaire was composed by 7 questions about stress, work engagement and frustration; 3 questions were daily self-administrated and 4 questions were weekly self-administrated.

2.2.2. Wearables Devices

The participants wore Xiaomi Mi Band 3 during 6 months. The activity, sleep and heart rate data were extracted by automatic data acquisition system of TALIONIS group, which was located in the work environment of participants.

3. Results

A total of 11 workers participated in this study, who were mostly women (63.63%). There were no significant differences between start and final measurements of participants.

The measures associated with health-related quality of life show that some participants reported to slight and moderate problems in the dimensions "pain/discomfort" (n = 2) and anxiety/depression (n = 3), specifically at the beginning of the study. In both evaluations, the average score of perceived health status were high (E_I = 84.56 (±8.13); E_F = 88.38 (±9.91)). PSS-10 results show that the men (Total score = 11.25) exhibited higher levels of stress than women (Total score = 9.71). This is opposed to weekly questionnaire, which shows that women (total score for women = 15.57; total score for man = 14.56) had higher level of stress.

Averages scores of PSQI show that participants had slight difficulties to fall asleep and low sleep quality (E_I = 4.89 (±2.62); E_F = 4.38 (±1.85)). By contrast, wearables reported that participants attained optimal sleep habits. Data from wearables Xiaomi Mi Band 3 show that men were sitting (7.35 h) and walking (2.22 h) more time than women (6.34 h; 1.64 h). However, the women slept for more time (8.15 h) than men (7.90 h).

4. Discussion and Conclusions

The aim of this proposal was to establish whether these wearables are devices capable of determining the level of occupational stress of workers. Furthermore, it was determined that the level of occupational stress and quality of life of a group of working people. EQ 5D-5L data show that there were no statically differences associated to sex; this matter is supported by other studies [8]. In general, the participants presented high levels of perceived health status. Men reported more stress than women; in contrast, scientific evidence has reported that the female sex had the highest levels of stress because social, psychological and biological consequences [9]. On the other hand, men showed fewer sleep hours than women; this aspect differs from others studies whose data indicated that women had more sleep problems than men [10]. In conclusion, it is necessary to develop studies with larger sample and with other working environments.

Author Contributions: Conceptualization, methodology, formal analysis and writing P.C.-M., M.C.M.-D., C.F. and B.G.; review, editing, visualization and supervision C.Q., A.J.P.S.M. and B.G.; and funding acquisition, M.C.M.-D. and P.C.-M. All authors have read and agreed to the published version of the manuscript.

Funding: The authors disclosed receipt of the following financial support for the research, authorship, and/or publication of this article: All the economic costs involved in the study will be borne by the research team. This work was supported in part by some grants from the European Social Fund 2014-2020. CITIC (Research Centre of the Galician University System) and the Galician University System (SUG) obtained funds through Regional Development Fund (ERDF) with 80%, Operational Programme ERDF Galicia 2014-2020 and the remaining 20% by the Secretaría Xeral de Universidades of the Galician University System (SUG). Specifically, the author PCM obtained a scholarship (Ref.ED481A-2019/069) and the author M.C.M.D. (Ref.ED481A 2018/205) to develop the PhD thesis. On the other hand, the diffusion and publication of this research was funded by the CITIC, Research Centre of the Galician University System with the support previously mentioned (Ref ED431G2019/01).

Acknowledgments: Financial support from the Xunta de Galicia and the European Union (European Social Fund—ESF), is gratefully acknowledged.

Conflicts of Interest: The authors declare no conflict of interest.

References

1. Galant-Miecznikowska, M.; Bhui, K.; Stansfeld, S.; Dinos, S.; de Jongh, B. Perceptions of work stress causes and effective interventions in employees working in public, private and non-governmental organisations: a qualitative study. *BJPsych Bull.* **2016**, *40*, 318–325.
2. Leka, S.; Griffiths, A.; Cox, T. La organización del trabajo y el estrés. *Protección la Salud los Trab.* **2004**, *3*, 1–37.
3. Maslach, C.; Schaufeli, W.B.; Leiter, M.P. Job Burnout. *Annu. Rev. Psychol.* **2001**, *52*, 397–422.
4. Estévez-Mujica, C.P.; Quintane, E. E-mail communication patterns and job burnout. *PLoS ONE* **2018**, 13, e0193966.
5. Neupert, S.D.; Almeida, D.M.; Mroczek, D.K.; Spiro, A. Daily stressors and memory failures in a naturalistic setting: Findings from the VA normative aging study. *Psychol. Aging* **2006**, *21*, 424–429.
6. Muaremi, A.; Arnrich, B.; Tröster, G. Towards Measuring Stress with Smartphones and Wearable Devices During Workday and Sleep. *Bionanoscience* **2013**, *3*, 172–183.
7. Han, L.; Zhang, Q.; Chen, X.; Zhan, Q.; Yang, T.; Zhao, Z. Detecting work-related stress with a wearable device. *Comput. Ind.* **2017**, *90*, 42–49.
8. Xie, F.; Pullenayegum, E.; Gaebel, K.; Bansback, N.; Bryan, S.; Ohinmaa, A.; Poissant, L.; Johnson, J.A. A Time Trade-off-derived Value Set of the EQ-5D-5L for Canada. *Med. Care* **2016**, *54*, 98–105.

9. Nordin, M.; Nordin, S. Psychometric evaluation and normative data of the Swedish version of the 10-item perceived stress scale. *Scand. J. Psychol.* **2013**, *54*, 502–507.
10. Uhlig, B.L.; Sand, T.; Ødegård, S.S.; Hagen, K. Prevalence and associated factors of DSM-V insomnia in Norway: The Nord-Trøndelag Health Study (HUNT 3). *Sleep Med.* **2014**, *15*, 708–713.

Proceedings

Virtual Reality Game Analysis for People with Functional Diversity: An Inclusive Perspective †

María del Carmen Miranda-Duro *, Patricia Concheiro-Moscoso, Javier Lagares Viqueira, Laura Nieto-Riveiro, Nereida Canosa Domínguez and Thais Pousada García

TALIONIS Group, CITIC, Oza, Universidade da Coruña, 15071 A Coruña, Spain; patricia.concheiro@udc.es (P.C.-M.); javier.lagares@udc.es (J.L.V.); laura.nieto@udc.es (L.N.-R.); nereida.canosa@udc.es (N.C.D.); thais.pousada.garcia@udc.es (T.P.G.)

* Correspondence: carmen.miranda@udc.es

† Presented at the 3rd XoveTIC Conference, A Coruña, Spain, 8–9 October 2020.

Published: 20 August 2020

Abstract: Virtual reality (VR) allows us to simulate everyday life environments with realism and in an immersive environment, with the use of the appropriate hardware. People with functional diversity, either because of environmental barriers or because of their reduced mobility, have fewer opportunities to participate in different daily activities or risk situations outdoors. Therefore, VR can be a technological resource for these people to access, try out, and experience different environments and scenarios, offering new participation experiences. Therefore, the aim of this proposal is to analyze the properties and determine the possibilities of the virtual reality applications available on commercial platforms for use in the practice of rehabilitation and intervention aimed at people with functional diversity. This is a transversal, descriptive study that has focused on the analysis of the 40 applications from the STEAM Virtual Reality and VIVE platforms for High Tech Computer Corporation (HTC). After analysis, it has been observed that there are no applications available that are fully accessible and with a minimum degree of usability for use by people with functional diversity.

Keywords: virtual reality; functional diversity; accessibility

---

1. Introduction

Virtual reality (VR) allows us to simulate everyday life environments with realism and in an immersive environment, with the use of the appropriate hardware. These virtual spaces allow for the reproduction of everyday situations, such as shopping, attending a concert, or going on a medical visit, but also, other experiences that are less frequent, such as exposure to risk events, adventure experiences, or stimuli that can simulate "phobias" for their approach. People with functional diversity, either because of environmental barriers or because of their reduced mobility, have fewer opportunities to participate in different daily activities or risk situations outdoors. Therefore, VR can be a technological resource for these people to access and try out different environments and scenarios, offering new experiences of participation. Furthermore, the possibility of adapting and customizing these virtual contents gives them added value, optimizing their applicability in a therapeutic intervention [1,2].

The aim of this proposal is to analyze the properties and determine the possibilities of VR apps available on commercial platforms for use in the practice of rehabilitation and intervention aimed at people with functional diversity. As a specific objective, it is proposed that the content of the apps must be related to different activities in people's daily lives, such as simulations of a job or activities such as cooking.

2. Material and Methods

The present study was carried out between February and May. The apps included in the analysis of the present study met the inclusion and exclusion previously defined (see Table 1).

The present study consists of a transversal, descriptive study that has focused on the analysis of the different apps included. This process has been carried out on the basis of a checklist of our own elaboration through which different characteristics were analyzed, such as language options; age for use; type of activity (observation or interaction); purpose; duration; physical and cognitive, comprehension or usability implications; game mode (individual or multiplayer); level of accessibility, and cost.

Table 1. Inclusion and exclusion criteria of virtual reality app analyzed.

Inclusion Criteria	Exclusion Criteria
Virtual Reality app about any activity of daily living (i.e., cooking, driving, job simulator)	Virtual Reality app about aggressive or conventional videogames
VR app from Steam VR and Viveport platform for HTC glasses	

3. Results

A total of 40 VR apps were included in the analysis of the present study. From Viveport were Camila, Let Hawaii Happen, Lifelique VR Museum, BallomBon Cashmere, Ikea VR Pancake Kitchen, The Stanford Ocean Acidification Experience, Richie Experience, Sim City, The Blu: Whale Encounter, VR Bowls, Engage, ISS 360º Tour with Tim Peake, Crossing the Road with Safety, Cmoar VR Cinema, Cabinet Model Room, The VR Museum of Fine Art, Underwater Mermaid, Clash of Chefs VR—Early Access, Bartender VR Simulator, Manifest Dream, VR Paris Bus Tour—France, Museum of Other Realities, 3D Organon VR Anatomy, OhShape, Ultimate Fishing Simulator, Mona Lisa: Beyond the Glass, First Person Tennis—The Real Tennis Simulator, Virtual Vacations, Blueplanet VR, Conductor, Paper Fire Rookie, and Healthy Badminton 2019 [3]. From the Steam platform, we included BoxVR, Tiny Town VR, Budget Cuts, Thief Simulator, Surgeon Simulator: Experience Reality, Ocean Rift, and Minigolf VR [4].

The most common language was English (92.5%), while the 52.5% of them were only in English, followed by Spanish (30%)—but none of them were only in Spanish—and Chinese (27.5%), while 7.5% were only in Chinese. The languages included in the apps were English, Spanish, Chinese, Korean, French, German, Italian, Japanese, Korean, Polish, Portuguese, Russian, Turkish, and Dutch. However, no app specified the appropriate age for use. In addition, related to the type of activity, 30% of the 40 apps were only for observational mode, while the rest were available in an observation and interaction mode with the VR scenario.

Among the different purposes that the different apps had, it is worth mentioning some of them, for example story-telling; vacation simulator; education (i.e., anatomy) and learning cultural aspects (i.e., visiting museums or cities); creating dresses; games such as blow up balloons; activities such as cooking; observation of oceans or landscapes; job simulator (i.e., office job, waiter, surgeon); a city simulator; playing sports (i.e., bowls, golf, tennis, badminton, boxing); organization of classes, meetings and others; space simulator in the role of an astronaut; crossing the road with safety; leisure activities such as cinema, room design, fishing, firefighting, and a thief simulator. The duration of the experience in each app was specified only in five of them and ranged from minimum 2 min to 2 hours maximum, depending on the type of activity. In addition, 82.5% are to be enjoyed individually, while the rest allows for a multi-player experience.

With regards to accessibility, 37.5% are required to be used standing up, while 5% are performed sitting down. However, 57.5% can be used both standing and sitting. In this line, 45% are free, while the rest range from EUR 0.99 to EUR 27.99.

4. Discussion and Conclusions

The aim of this proposal is to analyze the properties and determine the possibilities of VR apps available on commercial platforms for use in the practice of rehabilitation and intervention aimed at people with functional diversity. Therefore, it should be noted that the purpose of the apps mainly does not include in leisure activities, clothing, food, social participation, education, work, and play. Activities such as personal care, grooming, or others such as money management are also not included.

In the same way, the authors have found it difficult to classify the different apps as "easy", "difficult", or "intermediate" taking into account what is involved at the physical, cognitive or understanding level due to basic limitations, such as language or the controls themselves that are used for virtual reality experiences. Thus, there was no app specifically designed and created for people with functional diversity, but which also met the accessibility requirements defined by Web Content Accessibility Guidelines 2.1 [5].

In conclusion, after the analysis was carried out, it has been observed that there are no applications available that are fully accessible and with a minimum degree of usability for use by people with functional diversity. In addition, there is a language handicap, since they are mostly only available in English. The way of interacting with the virtual environment can also generate problems, since the use of the controls and the activation of different buttons complicates and limits the complex handling in case the person has reduced mobility. Given this lack of adequate resources and the relevance that VR can have in therapeutic processes, the need to create virtual reality scenarios adapted and suitable for people with functional diversity is detected.

Author Contributions: Conceptualization, methodology, formal analysis, and writing M.d.C.M.-D., P.C.-M., and J.L.V.; review, editing, visualization, and supervision, L.N.-R, N.C.D., and T.P.G.; funding acquisition, M.d.C.M.-D. and P.C.-M. All authors have read and agreed to the published version of the manuscript.

Funding: This work was supported in part by some grants from the European Social Fund 2014-2020. CITIC (Research Centre of the Galician University System) and the Galician University System (SUG) obtained funds through Regional Development Fund (ERDF) with 80%, Operational Programme ERDF Galicia 2014-2020 and the remaining 20% by the Secretaría Xeral de Universidades of the Galician University System (SUG). Specifically, the author MCMD obtained a scholarship (Ref. ED481A 2018/205) and the author PCM (Ref. ED481A-2019/069) to develop the PhD thesis. On the other hand, the diffusion and publication of this research was funded by the CITIC, Research Centre of the Galician University System with the support previously mentioned (Ref ED431G 2019/01).

Acknowledgments: Financial support from the Xunta de Galicia and the European Union (European Social Fund—ESF), is gratefully acknowledged.

Conflicts of Interest: The authors declare no conflict of interest.

References

1. Iberdrola: "Realidad Virtual: Otro Mundo al Alcance de Tus Ojos". Available online: https://www.iberdrola.com/innovacion/realidad-virtual (accessed on 15 June 2020).
2. Bote, M.A.; Martínez, A.L. Main theoretical approaches to functional diversity: A literature review. *Int. J. Disabil. Hum. Dev.* **2019**, *18*, 13–19.
3. Viveport. Available online: https://www.viveport.com/(accessed on 15 February 2020).
4. Steam. Available online: https://store.steampowered.com/steamvr?l=spanish (accessed on 15 February 2020).
5. W3C Web Accessibility Initiative WAI. Available online: https://www.w3.org/WAI/standards-guidelines/wcag (accessed on 15 February 2020).

Proceedings

Design and Implementation of a Physical Bitcoin Coin †

Alberto Femenias-Hermida [1,*], Cristian R. Munteanu [1,2] and José M. Vázquez-Naya [1,2]

1 Departamento de Computación, Facultad de Informática, Universidade da Coruña, Grupo RNASA-IMEDIR, Elviña, 15071 A Coruña, Spain; c.munteanu@udc.es (C.R.M.); jose@udc.es (J.M.V.-N.)
2 Centro de Investigación CITIC, Universidade da Coruña, Elviña, 15071 A Coruña, Spain
* Correspondence: alberto.femenias@udc.es
† Presented at the 3rd XoveTIC Conference, A Coruña, Spain, 8–9 October 2020.

Published: 20 August 2020

Abstract: One of the major factors hindering the adoption of crypto assets in general, and Bitcoin in particular, is the high level of complexity they present to the common user. Although physical coins are a possible solution, the need to place trust in the manufacturers (so that they throw away the private key) is a big drawback that has hampered their widespread use. The recent boom of the maker movement has brought in a significant number of users with access to 3D printing devices, as well as the supporting electronic and computing resources. We have taken advantage of these capabilities to develop an open source project that interested parties can use to easily print a physical model of a Bitcoin coin, along with the necessary software that allows the creation and validation of keys and addresses.

Keywords: bitcoin; open source coin; 3D printing; cryptographic asset

---

1. Introduction

To put it in an extremely simplified form, Bitcoin is a public ledger, stored in a single file that is shared with a p2p program, where participants keep their balances in a special form of accounts, called addresses. Each address, which is public, has an associated private key, that confers access to transfer the funds at will. The security of the funds relies in keeping the private keys safely stored and out of the reach of any malicious actor. The many instances were bitcoin owners have lost their funds [1] proves that keeping the private keys safe is a much harder problem than it may initially seem.

A possible solution is the use of a physical bitcoin, which is nothing but an artifact, whose shape and appearance resembles a traditional coin while containing in its interior the private key that gives its owner access to the associated funds. Their two main advantages are—first the use of the coin metaphor makes it very easy for regular users to visualize and identify them as money and second, their physical nature means that we can use the technologies developed over the course of centuries to keep them secure. However, a significant concern associated with physical bitcoin coins is the need to entrust the manufacturer with the disposal of the private key upon creation of each coin. That requires a big leap of faith, which has probably kept physical bitcoins from achieving a much greater level of adoption. In this paper, we present a solution that solves that problem, by allowing end users to create their own physical bitcoins without having to trust any third party.

2. Materials and Methods

We have used Github to make publicly available the results of our work, namely: (a) The software, written in Python (b) The CAD model (created with Autodesk Fusion 360) and (c) All the associated documentation, both for the process of creating the coin, as well as for its ulterior use.

The repository can be accessed at: https://github.com/albertofemenias/bertocoin.

In order to reproduce all the results of this project, the user will need:

(1) An additive 3D printer with PLA filament, such as a Prusa or similar,
(2) A good quality paper printer, such as a generic laser or inkjet printer,
(3) A transparent plastic laminate and a metallic washer.

The software and the 3D models were developed using an incremental process under the Kanban methodology. We rely on the benefits of the open source model to ensure that the design is publicly audited so that the users can trust the design is secure and free of malicious code.

3. Development

After reviewing the state of the art regarding physical bitcoins, we started the project by studying the possibilities of creating a sound design, that could be easily manufactured in an ordinary, maker class 3D printer. Of particular significance is the issue of making a functional seal. The seal is a physical mechanism of the coin with two main properties: (a) the user must manipulate it in order to gain access to the private key of the coin and (b) once the seal has been activated (typically broken) it must be obvious thereafter that the security of the coin has been compromised.

We approached an iterative design with several rounds of trial and error until we finally arrived at a design that we deemed good enough. The final design makes use of a mesh of low-caliber, parallel strands of material that the user must break to access the private key stored internally. The act of breaking the mesh and folding the flap to access the contents of the coin inflicts a permanent and easily visible damage that clearly indicates that the private key has been revealed.

4. Results

Special attention was given to make the software as user-friendly as possible. As a result, it consists of a single Python script that, when run, generates the keys in a secure manner and produces automatically a printable document (in the open source .SVG format) that the user can print using an internet browser. This document (see Figure 1) contains both the private and the public key, and it is designed so that it can be easily cut out and folded, before placing it inside the coin.

Figure 1. Template with private and public keys.

As for the physical side of the coin, most of the effort was placed on three critical aspects—(a) Making the coin easily recognizable as such. This required finding an appropriate shape, look, weight and size; (b) Designing a workable seal, that makes pretty evident the fact that it has been broken to allow access to the secret key; and c) Guaranteeing the secret is not accessible without breaking the seal. This was ensured in good part by inserting a metallic washer inside the coin that makes the coin 100% opaque. In Figure 2, we can see various aspects of the design and manufacturing of the coin.

Figure 2. Design and manufacturing of the coin.

5. Discussion and Conclusions

We have created a workable solution to handle bitcoin funds as a coin. The design will surely benefit from public scrutiny, once enough qualified people analyze the code published in the Github repository.

It is critical that the users of these coins understand two key aspects of their security: (a) They must be produced in a non-compromised computer to ensure a fair private key is generated and (b) Upon manufacturing, the coins must be physically kept out of reach of anybody but the legitimate user, since they are essentially 'bearer assets' which means that whoever is in possession of them can access the associated bitcoin funds.

6. Future Work

Upon finishing and testing the design, we identified a line of work for the future that involves embedding electronics within the coin, to improve some of its characteristics. Most notably the incorporation of computing and I/O capabilities in the coin will make it interactive, enabling non-destructive verification of the private key, by building upon the cryptographical properties of the ECC to sign a message with the private key.

Author Contributions: Conceptualization, A.F.-H.; Methodology, A.F.-H., C.R.M. and J.M.V.-N.; Software, A.F.-H.; Investigation, A.F.-H., C.R.M. and J.M.V.-N.; Resources, C.R.M. and J.M.V.-N.; Writing—original draft preparation, A.F.-H.; Writing—review and editing, A.F.-H., C.R.M. and J.M.V.-N.; Supervision, C.R.M. and J.M.V.-N. All authors have read and agreed to the published version of the manuscript.

Funding: This work was supported by the Consolidation and Structuring of Competitive Research Units—Competitive Reference Groups (ED431C 2018/49), funded by the Ministry of Education, University and Vocational Training of the Xunta de Galicia endowed with EU FEDER funds.

Conflicts of Interest: The authors declare no conflict of interest.

Reference

1. Eriksson, N. 10 Dramatic Stories of People Who Lost Their Bitcoin Private Keys. *Coinnounce*. Available online: https://coinnounce.com/10-dramatic-stories-of-people-who-lost-their-bitcoin-private-keys/ (accessed on 22 July 2020).

Proceedings

An Agent-Based Model to Simulate the Spread of a Virus Based on Social Behavior and Containment Measures †

Manuel Seijas Carpente *,‡, Bertha Guijarro-Berdiñas ‡, Amparo Alonso-Betanzos ‡, Alejandro Rodríguez-Arias ‡ and Adina Dimitru ‡

University of A Coruña, CITIC, Campus de Elviña, s/n, 15071 A Coruña, Spain; berta.guijarro@udc.es (B.G.-B.); amparo.alonso@udc.es (A.A.-B.); alejandro.rodríguez.arias@udc.es (A.R.-A.); adina.dumitru@udc.es (A.D.)

* Correspondence: manuel.seijas@udc.es; Tel.: +34-655-29-96-27

† Presented at the 3rd XoveTIC Conference, A Coruña, Spain, 8–9 October 2020.

‡ These authors contributed equally to this work.

Published: 20 August 2020

Abstract: COVID-19 has brought a new normality in society. However, to avoid the situation, the virus must be stopped. There are several ways in which the governments of the world have taken action, from small measures like general cleaning up to large-scale measures like confinement. In this work, we present an agent-based tool that allows for simulating the virus expansion as a function of these containment measures and the Social Behavior based on people needs, beliefs, and social relations. Once this tool has been validated, it will be useful to evaluate the impact of future containment measures so that the most balanced ones can be found for the effectiveness of the measures and their good reception by the population.

Keywords: intelligent software agents; agent-based modeling; artificial intelligence; NetLogo; COVID-19; epidemiological model

---

1. Introduction

The impact of COVID-19 on society has generated important changes in it. Countries have been forced to take different control measures with different responses and results in their inhabitants. In this situation, it becomes mandatory to answer questions like: How will people's behavior change in the face of the measures taken? How will the rate of spread of the virus change depending on people's behavior?

Taking all these questions into consideration, we want to reach a solution that benefits both the well-being of society (including its satisfaction and needs) as well as the containment of the virus and danger. To achieve the proposed solution, it is important to be able to simulate all kinds of situations in which not only the preventive measures and actions in which governments agree to reduce expansion are taken into account, but also the variables of social satisfaction, as well as social needs.

2. The Agent-Based Model

In this section of the paper, we will talk about the project itself, its phases, results, and conclusions.

2.1. Phase 1: Virus Spread Model

In the first phase, the objective was to introduce a basic model into the system that would allow for simulating the spread of the virus in humans, taking into account exclusively their social relationships (friends, family, close people, etc.). None of the containment measures are taken into account at this stage. To carry this out, we chose the SIR epidemiological model [1], which represents with almost total precision the real situation. However, this model needs certain adjustments, to add social relations between individuals, as follows:

- The related individuals are more likely to infect each other than to/or other individuals.
- Individuals in a certain network of relationships can infect people outside that network or be infected by them.
- Individuals are more likely to have contact with people in their network, but they can also have contact with people outside of that network.

These rules may seem a little trivial, but they must be defined to create something akin to a real environment in which the virus spreads.

As a development tool, we use NetLogo, a multi-agent programmable modeling environment. We provide the user with an interface in which this network is defined with a color code: green individuals simulate people who may be infected (they are not but can be infected by others), red individuals simulate infected people, and blue individuals simulate cure/dead people (in this phase, it is not necessary to know the difference). Individuals can change between states; a green individual can turn red if infected, a red can turn blue if he cures, and so on.

2.2. Phase 2: Modeling Human Behavior

When governments have faced the problem of containing the COVID-19 contagion, in almost all cases, regulations have been issued that involve sudden changes in people's habits. Even if these rules are dictated to change a situation for the better and for the general benefit of society, they affect people's lives differently. Based on sociological and psychological research, we endowed agents with certain behaviors that make them react differently to the same norms. To do this, we use the HUMAT model [2] in which agents behave according to values (like environmental care), social needs (like feeling accepted by a group), and experiential needs (like the need to go for a walk or earn money). These needs, and how they are satisfied, can cause dilemmas or cognitive dissonance in agents (for example, an agent's values force him to obey the rules, but he has an urgent need to go outside). Finally, the strength of these cognitive dissonances is what will cause the agent to break government-mandated rules or even try to convince others to break them.

2.3. Phase 3: Data Gathering

The next step was to collect information about people's thoughts and feelings about possible virus containment measures in Spain. In addition to this, we collected data on their way of life and their possibilities in different situations, to name a few:

- In a quarantine situation, how they would feel, considering their salary, the size of their home, the number of children, home situations, etc.
- Other questions focused on people who, regarding the type of measures taken, have to keep working and pushing themselves to the limit of contagion, just to keep the economy somewhat stable.
- Finally, in this phase, information was also obtained on the psychological aspects of living under certain potential future control measures.

All of this data, and more, is useful for generating simulations in which the artificial human society can take certain actions or decisions that would lead the same society to different results in the spread of the virus (for example, people who tend to skip rules because they are living an uncomfortable situation are more likely to massively spread contagions).

2.4. Phase 4: Near Future Work, Applying the Data to the Sir Model

The last step will be to use the gathered data to improve the virus expansion model. In this phase, our individuals will be able to make decisions and take actions on their social network, such as breaking rules or making other people break them as well.

Historical data and survey data will help us to reconstruct the lived situation and simulate others in which a a community succeeds or fails to follow social norms and stop contagions.

With all of this information, the system will be updated into a AI System where the information will lead the virus in its expansion, obtaining different results. These results can be measured and interpreted, searching for the best strategy to follow on a situation like the one of the simulation.

3. Conclusions

An agent-based model has been developed that includes a classical model of virus expansion that is further modified thanks to the ability of the system to simulate the behavior of a society with respect to certain norms. This system will be adapted to the COVID situation by using data gathered from surveys on April 2020 and by including the political norms the the Spanish Government implemented to stop contagions. Such an agent-based simulation of a case will help explain what social dynamics played a role and how critical actions and needs of people affected these dynamics. In this way, we can learn the conditions under which better or worse scenarios could have been more probable and help to find a balance between the effectiveness of the measures and their good reception by the population.

Funding: This research received no external funding.

Conflicts of Interest: The authors declare no conflict of interest.

References

1. Jager, W.; Janssen, M.A. An updated conceptual framework for integrated modeling of human decision making: The Consumat II. *Psychology* **2012**.
2. Bacaër, N. McKendrick and Kermack on epidemic modelling (1926–1927). In *A Short History of Mathematical Population Dynamics*; Springer: London, UK, 2011; pp. 89–96.

Proceedings

Shiny Dashboard for Monitoring the COVID-19 Pandemic in Spain †

Carlos Fernandez-Lozano [1,2,*] and Francisco Cedron [1,2]

1 Department of Computer Science and Information Technologies, Faculty of Computer Science, Universidade da Coruña, Campus Elviña s/n, 15071 A Coruña, Spain; francisco.cedron@udc.es

2 CITIC-Research Center of Information and Communication Technologies, Universidade da Coruña, 15071 A Coruña, Spain

* Correspondence: carlos.fernandez@udc.es; Tel.: +34-881-01-6013

† Presented at the 3rd XoveTIC Conference, A Coruña, Spain, 8–9 October 2020.

Published: 20 August 2020

Abstract: Real-time monitoring of events such as the recent pandemic caused by COVID-19, as well as the visualization of the effects produced by its expansion, has highlighted the need to join forces in fields already widely used to working hand in hand, such as medicine, biology and information technology. Our dashboard is developed in R and is supported by the Shiny package to generate an attractive visualization tool: COVID-19 Spain automatically produces daily updates from official sources (Carlos III Research Institute and Ministry of Health, Consumer Affairs and Welfare) in cases, deaths, recovered, ICU admissions and accumulated daily incidence. In addition, it shows on a georeferenced map the evolution of active, new and accumulated cases by autonomous community allowing to travel in time from the origin to the last available day, which allows to visualize the expansion of infections and serves as a visual support for epidemiological studies.

Keywords: COVID-19; R; Shiny; monitoring

---

1. Introduction

The pandemic generated by COVID-19 has highlighted concepts such as reproducibility or interactive publication of results for real-time monitoring of an event, two of the battlefields of research today where most of the models proposed in different scientific publications are not easily accessible, analyzable or reproducible. There is no doubt that since the end of 2019, once it became known that uncontrolled outbreaks of infection by a coronavirus were occurring and were severely affecting the respiratory system, the interest at a global level in obtaining information on the monitoring of this infection led to the search for multiple research groups for methods/systems that, in addition to predicting the evolution, would allow the evolution of the pandemic to be visualized as simply as possible. As of 20 July 2020, the World Health Organization (WHO) has records of more than 14.3 million confirmed cases and more than 600,000 deaths, of which more than 260,000 confirmed cases correspond to Spain and more than 28,000 deaths. The data are so overwhelming that they undoubtedly show the seriousness of the pandemic that has been experienced worldwide. In Spain, for example, the situation of saturation of the health systems led in mid-March to the declaration of a State of Alarm throughout the country, preventing the free movement of citizens and closing borders. It is precisely this situation that led to the development of a Dashboard using R [1] and Shiny [2].

2. Results

As previously mentioned the website is available at https://covid19.citic.udc.es and was created with the initial objective of serving as a visual aid to the spread of the pandemic in Spanish territory.

To this end, the dashboard has a dynamic and interactive general map of the country's evolution (Figure 1) which allows the data to be displayed from the beginning of the series.

Figure 1. Overview of the shiny dashboard.

In the dashboard you can analyze the day-to-day evolution of the pandemic in Spain by Autonomous Community and variable of interest (Figure 2a) or the degree of infection by population pyramid (Figure 2b) using the ggplot2 [3] and plotly [4] libraries.

(**a**) (**b**)

Figure 2. Interactive pandemic progress charts. (**a**) Accumulated cases (log10); (**b**) Confirmed counts by age range.

The deployment was carried out on a docker container with a swarm orchestrator in charge of balancing the load to avoid dashboard saturation in the face of a high number of simultaneous accesses.

3. Conclusions

A dynamic and interactive web dashboard has been developed to visualize the evolution of COVID-19 infection at a national level also segregating by autonomous community.

Funding: This project is supported by the General Directorate of Culture, Education and University Management of Xunta de Galicia (Ref. ED431G/01, ED431D 2017/16), Competitive Reference Groups (Ref. ED431C 2018/49).

Acknowledgments: Technical support from CITIC and UDC, specially to Carlos J. Escudero and Alejandro Mosteiro.

References

1. R Core Team. *R: A Language and Environment for Statistical Computing*; R Foundation for Statistical Computing: Vienna, Austria, 2020.
2. Chang, W.; Cheng, J.; Allaire, J.; Xie, Y.; McPherson, J. *Shiny: Web Application Framework for R*; R Package Version 1.4.0.2; https://cran.r-project.org/web/packages/shiny/index.html; 2020.

3. Wickham, H. *Ggplot2: Elegant Graphics for Data Analysis;* Springer: New York, NY, USA, 2016.
4. Sievert, C. *Interactive Web-Based Data Visualization with R, Plotly, and Shiny;* Chapman and Hall/CRC: London, UK, 2020.

Proceedings

Decentralized P2P Broker for M2M and IoT Applications †

Iván Froiz-Míguez [1,2], Paula Fraga-Lamas [1,2,*] and Tiago M. Fernández-Caramés [1,2,*]

1 Department of Computer Engineering, Faculty of Computer Science, Universidade da Coruña, 15071 A Coruña, Spain; ivan.froiz@udc.es
2 CITIC Research Center, Universidade da Coruña, 15071 A Coruña, Spain
* Correspondence: paula.fraga@udc.es (P.F.-L.); tiago.fernandez@udc.es (T.M.F.-C.); Tel.: +34-981-167-000 (ext. 6051) (P.F.-L.)
† Presented at the 3rd XoveTIC Conference, A Coruña, Spain, 8–9 October 2020

Published: 20 August 2020

Abstract: The recent increase in the number of connected IoT devices, as well as the heterogeneity of the environments where they are deployed, has derived into the growth of the complexity of Machine-to-Machine (M2M) communication protocols and technologies. In addition, the hardware used by IoT devices has become more powerful and efficient. Such enhancements have made it possible to implement novel decentralized computing architectures like the ones based on edge computing, which offload part of the central server processing by using multiple distributed low-power nodes. In order to ease the deployment and synchronization of decentralized edge computing nodes, this paper describes an M2M distributed protocol based on Peer-to-Peer (P2P) communications that can be executed on low-power ARM devices. In addition, this paper proposes to make use of brokerless communications by using a distributed publication/subscription protocol. Thanks to the fact that information is stored in a distributed way among the nodes of the swarm and since each node can implement a specific access control system, the proposed system is able to make use of write access mechanisms and encryption for the stored data so that the rest of the nodes cannot access sensitive information. In order to test the feasibility of the proposed approach, a comparison with an Message-Queuing Telemetry Transport (MQTT) based architecture is performed in terms of latency, network consumption and performance.

Keywords: IoT; edge computing; M2M; distributed computing; IPFS; MQTT; P2P

1. Introduction

The growing number of Internet of Things (IoT) devices generates a massive amount of data that has derived into the adoption of different communications paradigms that go beyond traditional client-server schemes. One of such paradigms is edge computing [1], which is based on the distribution of the computing load among different IoT nodes, thus moving part of the processing from the cloud to the edge of the network and then providing lower latency, improved response times and better bandwidth availability. In addition, recent advances on hardware enable creating more powerful and less power-hungry devices. Thus, the latest IoT devices can handle more complex tasks than simple data storage and device-to-device communications.

The mentioned evolution fosters the development of new distributed computing strategies like the one described in this paper. The proposed strategy is completely distributed, in contrast to traditional edge computing approaches, which provide a hybrid environment with distributed computing and a central server (i.e., a cloud). Not delegating information to a central server has become increasingly important, as it is a single point of failure and a potential source of data leaks. Therefore, the proposed solution makes use of a fully distributed Machine-to-Machine (M2M) communications protocol whose performance is compared with Message Queuing Telemetry Transport (MQTT) [2], which is currently one of the most popular M2M protocols.

2. Design and Implementation

Figure 1 shows the proposed communications architecture. In such an architecture a private swarm is a set of peers that belong to the IoT system. Each peer is a device that manages the communications distributed among the different sensor nodes. Every peer provides persistent storage for the data gathered from its edge computing-based network. The communication between a user and the edge computing-based network is carried out within the same Local Area Network (LAN) through a REST API.

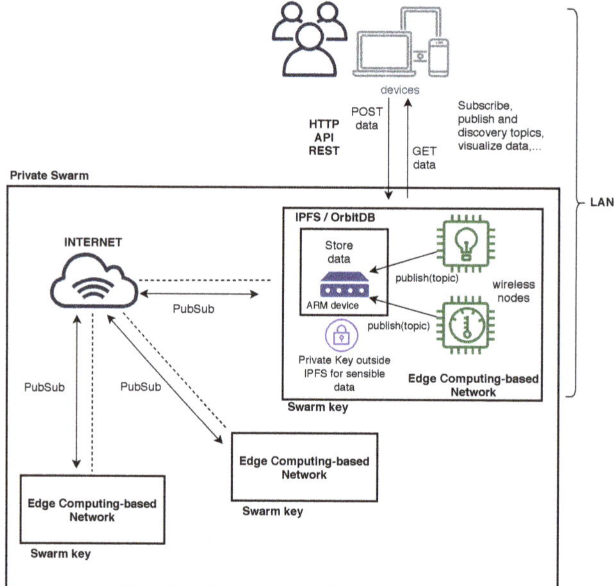

Figure 1. Communications architecture of the proposed system.

The devices make use of Inter-Planetary File System (IPFS) [3] to implement a decentralized file system that provides better performance than HTTP when managing large amounts of data. Moreover, the devices use a experimental publication/subscription protocol called PubSub for M2M communications.

For implementing the proposed architecture, a Raspberry Pi was used as a node. It runs a go-ipfs instance with the PubSub function enabled. For persistent storage, OrbitDB (a distributed database that runs on top of IPFS) is used through a port in golang that offers better performance than the original javascript version [4]. In addition, it is possible to communicate with the system via an HTTP REST API to perform actions and get the obtained results.

As edge device, a Raspberry Pi Zero (RPi Zero) was used to receive measurements from different sensors via BLE or WiFi. Thus, the RPi Zero is in charge of the distributed storage and of the M2M communications with other edge devices.

3. Experiments

To determine the performance of the system in terms of latency and throughput, different tests were carried out. The obtained results were compared with the ones provided by an MQTT broker that run in a cloud (an Eclipse Mosquitto broker was deployed in a cloud while a client node (RPi Zero) published messages). In a similar way, an IPFS node hosted in the same cloud acted as a topic subscriber while the client node sent messages. The tests simulated the publication of 10 messages from 10 different clients. The obtained latencies are shown in Table 1.

Table 1. Latency comparison between MQTT (left) and IPFS PubSub (right).

Measure	Latency (ms)	Measure	Latency (ms)
Minimum publish time	41.101	Minimum publish time	320.8
Maximum publish time	69.067	Maximum publish time	375.539
Mean publish time	52.379	Mean publish time	340.272
Standard deviation	5.649	Standard deviation	6.173

In addition, the throughput of OrbitDB was measured by making insertions in an EventLog and then measuring response times. Table 2 shows the obtained results.

Table 2. Performance of an EventLog of OrbitDB.

Number of Insertions	Latency (s)	Throughput (queries/s)
50	7.558	6.615
100	12.777	7.826
500	72.82	6.866

4. Conclusions

The proposed decentralized brokerless system offers a good trade-off between performance, security, and reliability. Although MQTT provides a low latency (maintly beacause PubSub was not designed by having M2M communications in mind), its centralized architecture is prone to security issues that can be easily tackled by the proposed system.

References

1. Shi, W.; Cao, J.; Zhang, Q.; Li, Y.; Xu, L. Edge Computing: Vision and Challenges. *IEEE Internet Things J.* **2016**, 3, 637–646.
2. Naik, N. Choice of effective messaging protocols for IoT systems: MQTT, CoAP, AMQP and HTTP. In Proceedings of the 2017 IEEE international systems engineering symposium (ISSE), Vienna, Austria, 11–13 October 2017.
3. Benet, J. IPFS—Content Addressed, Versioned, P2P File System. IPFS White-Paper. *arXiv* **2014**, arXiv:1407.3561.
4. Froiz-Míguez, I.; Fraga-Lamas, P.; Varela-Barbeito, J.; Fernández-Caramés, T.M. LoRaWAN and Blockchain based Safety and HealthMonitoring System for Industry 4.0 Operators. *Proceedings* **2019**, 42, 77.

Proceedings

Joint Optic Disc and Cup Segmentation Using Self-Supervised Multimodal Reconstruction Pre-Training †

Álvaro S. Hervella [1,2,*], Lucía Ramos [1,2], José Rouco [1,2], Jorge Novo [1,2] and Marcos Ortega [1,2]

1 Centro de Investigación CITIC, Universidade da Coruña, 15071 A Coruña, Spain; l.ramos@udc.es (L.R.); jrouco@udc.es (J.R); jnovo@udc.es (J.N.); mortega@udc.es (M.O.)
2 VARPA Research Group, Instituto de Investigación Biomédica de A Coruña (INIBIC), Universidade da Coruña, 15006 A Coruña, Spain
* Correspondence: a.suarezh@udc.es; Tel.: +34-981-167-000
† Presented at the 3rd XoveTIC Conference, A Coruña, Spain, 8–9 October 2020.

Published: 20 August 2020

Abstract: The analysis of the optic disc and cup in retinal images is important for the early diagnosis of glaucoma. In order to improve the joint segmentation of these relevant retinal structures, we propose a novel approach applying the self-supervised multimodal reconstruction of retinal images as pre-training for deep neural networks. The proposed approach is evaluated on different public datasets. The obtained results indicate that the self-supervised multimodal reconstruction pre-training improves the performance of the segmentation. Thus, the proposed approach presents a great potential for also improving the interpretable diagnosis of glaucoma.

Keywords: deep learning; self-supervised learning; segmentation; eye fundus; glaucoma

---

1. Introduction

The detailed analysis of the optic disc and optic cup in retinal images is a key step for the diagnosis of glaucoma. In this regard, several biomarkers derived from the morphological analysis of these two structures have demonstrated to be useful for the diagnosis and screening of the disease. This has motivated the development of automated methods for the segmentation of optic disc and cup in retinography. Nowadays, the most successful segmentation approaches are those based on deep learning techniques. However, the training of deep neural networks requires large amounts of annotated data that can be difficult to obtain.

In this work, we present a novel approach for the joint segmentation of the optic disc and cup in retinography using deep learning [1]. Given the limited size of common datasets, we propose a novel self-supervised pre-training consisting in a multimodal reconstruction between complementary retinal image modalities. To validate this proposal, we perform experiments on different public datasets.

2. Methodology

The proposed pre-training consists of the self-supervised multimodal reconstruction of fluorescein angiography from retinography. This multimodal reconstruction leverages the availability of unlabeled multimodal image pairs for learning about the retinal anatomy [2]. In particular, the self-supervised multimodal reconstruction is trained as proposed in Reference [3] using a public dataset of retinography-angiography pairs. After the pre-training phase, the neural network is fine-tuned in the target task, that is, the joint segmentation of the optic disc and cup in retinography. This joint segmentation is approached as a pixel-wise multi-class classification where the neural network learns to predict the likelihood of the different classes [1].

3. Results and Conclusions

The proposed methodology is evaluated on two different public datasets, which present a representative variety of glaucomatous an healthy retinas. Additionally, we also perform a comparison against training the network from scratch in the segmentation task, which is the most common approach in the literature. The quantitative evaluation shows that the proposed methodology significantly improves the segmentation of the optic disc and optic cup. In particular, our proposal achieves a Jaccard index of 82.29 and 92.43 for the optic cup and disc, respectively, whereas the training from scratch achieves a Jaccard index of 75.36 and 88.19 for the optic cup and disc, respectively. Additionally, in comparison to other alternatives in the literature, our proposal achieves a competitive state-of-the-art performance while using less annotated data. Figure 1 depicts a representative example of predicted segmentations, where it is observed that the proposed approach produces more consistent segmentations than the alternative training from scratch. Conclusively, the self-supervised multimodal reconstruction pre-training demonstrates to be useful for improving the joint segmentation of the optic disc and cup in retinography.

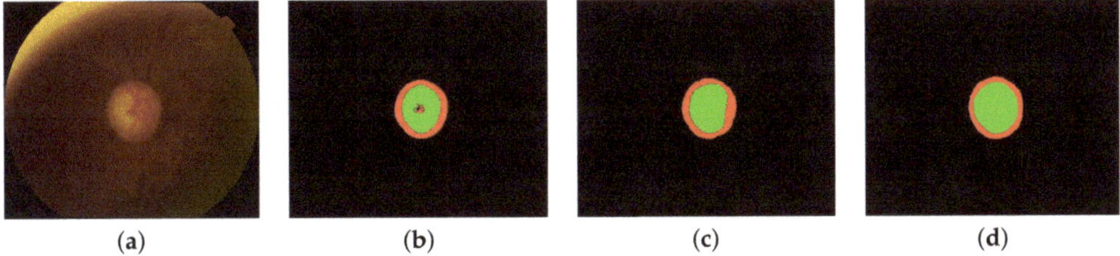

(a) (b) (c) (d)

Figure 1. Representative example of predicted segmentations for a glaucomatous retina. (**a**) Retinography, (**b**,**c**) predicted segmentations, and (**d**) ground truth segmentation. The predicted segmentations correspond to (b) a neural network trained from scratch and (c) the proposed approach.

Author Contributions: A.S.H., J.R. and J.N. contributed to the analysis and design of the computer methods and the experimental evaluation methods, whereas A.S.H. also developed the software and performed the experiments. L.R and M.O. contributed with domain-specific knowledge. All the authors performed the result analysis. A.S.H. was in charge of writing the manuscript, and all the authors participated in its critical revision and final approval. All authors have read and agreed to the published version of the manuscript.

Acknowledgments: This work is supported by Instituto de Salud Carlos III, Government of Spain, and the European Regional Development Fund (ERDF) of the European Union (EU) through the DTS18/00136 research project, and by Ministerio de Ciencia, Innovación y Universidades, Government of Spain, through the RTI2018-095894-B-I00 research project. The authors of this work also receive financial support from the ERDF, the European Social Fund (ESF) of the EU, and Xunta de Galicia through Centro de Investigación de Galicia ref. ED431G 2019/01 and the predoctoral grant contract ref. ED481A-2017/328.

Conflicts of Interest: The authors declare no conflict of interest.

References

1. Hervella, Á.S.; Ramos, L.; Rouco, J.; Novo, J.; Ortega, M. Multi-Modal Self-Supervised Pre-Training for Joint Optic Disc and Cup Segmentation in Eye Fundus Images. In Proceedings of the 2020 IEEE International Conference on Acoustics, Speech and Signal Processing (ICASSP), Barcelona, Spain, 4–8 May 2020; pp. 961–965, doi:10.1109/ICASSP40776.2020.9053551.
2. Hervella, Á.S.; Rouco, J.; Novo, J.; Ortega, M. Learning the retinal anatomy from scarce annotated data using self-supervised multimodal reconstruction. *Appl. Soft Comput.* **2020**, *91*, 106210, doi:10.1016/j.asoc.2020.106210.
3. Hervella, Á.S.; Rouco, J.; Novo, J.; Ortega, M. Self-supervised multimodal reconstruction of retinal images over paired datasets. *Expert Syst. Appl.* **2020**, *161*, 113674, doi:10.1016/j.eswa.2020.113674.

Proceedings

Analysis and Definition of Data Flows Generated by Bio Stimuli in the Design of Interactive Immersive Environments †

Paulo Veloso Gomes [1,*], João Donga [2], António Marques [1], João Azevedo [2] and Javier Pereira [3]

1 LabRP, Psychosocial Rehabilitation Laboratory, School of Allied Health Technologies, Polytechnic Institute of Porto, 4200-072 Porto, Portugal; ajmarques@ess.ipp.pt
2 LabRP, Psychosocial Rehabilitation Laboratory, School of Media Arts and Design, Polytechnic Institute of Porto, 4200-072 Porto, Portugal; jpd@esmad.ipp.pt (J.D.); joaoazevedo@esmad.ipp.pt (J.A.)
3 CITIC-Research Center of Information and Communication Technologies, University of A Coruña, 15071 A Coruña, Spain; javier.pereira@udc.es
* Correspondence: pvg@ess.ipp.pt

† Presented at the 3rd XoveTIC Congress, A Coruña, Spain, 8–9 October 2020.

Published: 20 August 2020

Abstract: This work focuses on interactivity as one of the essential factors for creating immersive environments, particularly interactivity that generates involuntary responses over which the user does not have conscious control. A dynamic and adaptive model was designed to analyze and define the data flow generated by bio stimuli for the design of interactive immersive environments.

Keywords: mental health and wellness; affective computing; empathy; immersive environments; augmented reality; virtual reality; electroencephalography; biofeedback; affective feedback

---

1. Introduction

Immersive environments provide impactful experiences that generate different types of emotions. The more immersive the environment, the greater influence it will have on the user's perception of their involvement with that environment, creating a sense of realism and a sense of presence in it. The feeling of immersion is caused by a set of factors. Interactivity is based on the emission of stimuli, which can be of different types, inducing responses by the user. In the process of the interactivity of an immersive system, stimuli intend to trigger responses.

Responses to stimuli can be voluntary, when the user is aware of the response they intend to give and choose to respond to the stimulus in a certain way, or involuntary when the user does not control the response. However, self-control of a person's biological status can be developed with time and experience [1]. This biofeedback process can control physiological factors such as heart rate and brain waves, among others [1].

One of the three core areas of affective computing that provide relevant methods and techniques to affective design is related with emotion sensing and recognition [2]. The concept of affective feedback relates the concepts of affective computing and biofeedback, and its application in immersive environments intends that the use of biofeedback mechanisms in the system will influence the user experience [1].

2. Objectives

The objective of this work is to design a dynamic and adaptive model to analyze and define the data flow generated by bio stimuli for the design of interactive immersive environments.

1. Identify the bio-response inducing stimuli sent by the system.
2. Detect, analyze, and classify the response by the user using biofeedback mechanisms.
3. Convert the obtained values into a scale and relate them to inducing stimuli.
4. Use the values obtained by the biofeedback to redefine the system parameters and emit new stimuli.

3. Methods

Immersive environments are designed to affect the user; exposure causes involuntary reactions, such as changes in the heart and respiratory rate and changes in the electrical brain activity and eye movements. These reactions can be captured in real time by biofeedback devices capable of identifying these changes and measuring their intensity. The real-time data obtained on the user's physiological aspects allows us to determine how the stimuli affect them (Figure 1). On the other hand, when the user receives information in real time about a certain aspect of his physiology, they can determine how their mental changes can influence their state.

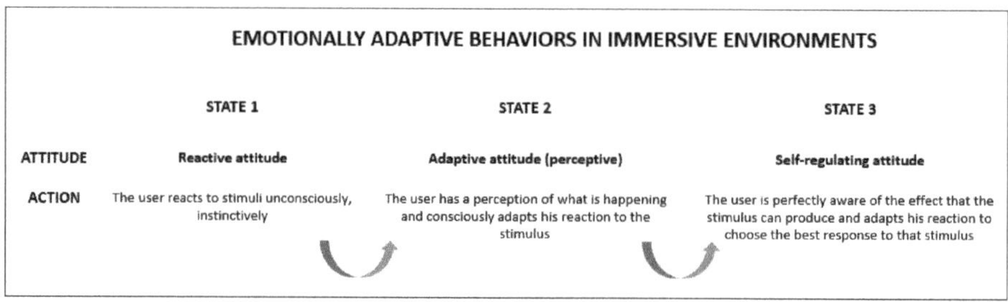

Figure 1. Behavioral change during continuous exposure to stimuli generated in immersive environments.

To optimize the user experience, an emotionally adaptive system continuously adapts its stimulus to the user's emotional state, through the measurement of their emotional data [3]. To incorporate the effects of stimuli on the interactivity process, the system first processes the unimodal data separately and then merges them [4] in order to interpret the user's response and use that response to generate new stimuli.

The perception and externalization of emotions can occur voluntarily and involuntarily. The participant can feel one thing and convey that they felt another. The use of multimodal data increases the reliability of the system by detecting different sources of biological signals [5].

The introduction of biofeedback systems in the design of a simple immersive environment transform it into an Emotionally Adaptive Immersive Environment, where the user experience can be optimized through the continuously adaptation of stimuli to the user emotional state. The quantity and intensity of the stimuli are determined through an adaptive affective algorithm which collects, interprets, and converts the user's physiological data.

To explore the potential of interactivity in immersive environments, the simple stimulus-response model must evolve into a dynamic and adaptive model. This model incorporates a set of cycles between stimuli and responses.

The system emits stimuli that trigger physiological reactions; the data generated by these reactions is sent and interpreted by the algorithm, which identifies the type and intensity of the emotion created by the stimulus. The affective algorithm adapts the system's responsiveness through mapping between the data obtained, considering the type and intensity of the stimuli, with the appropriate response to each situation, selecting new stimuli to emit and grading the respective intensity.

4. Discussion/Conclusions

Since interactivity is an important factor in the construction of immersive environments, the design of a dynamic and adaptive model to analyze and define the data flow generated by bio stimuli is fundamental for the conceptualization of the system.

The Emotionally Adaptive Immersive Environment through an affective algorithm can use the Not Intentional Mode Strategy (NIMS), based on randomness in the stimulus management process, or the Intentional Mode Strategy (IMS), which automatically generates stimuli according to pre-defined objectives, but it also allows the Controlled Mode Strategy (CMS) through the intervention of a supervisor who controls the system during user exposure to the immersive system.

Author Contributions: Conceptualization, P.V.G.; methodology, P.V.G. and A.M.; validation, P.V.G., J.D., and J.A.; investigation, P.V.G.; writing—original draft preparation, P.V.G.; writing—review and editing, P.V.G.; visualization, P.V.G and J.D.; supervision, A.M. and e.J.P.; project administration, P.V.G.

Funding: This research received no external funding.

Acknowledgments: This research was carried out and used the equipment of the Psychosocial Rehabilitation Laboratory (LabRp) of the Research Center in Rehabilitation of the School of Allied Health Technologies, Polytechnic Institute of Porto.

Conflicts of Interest: The authors declare no conflict of interest.

References

1. Bersak, D.; McDarby, G.; Augenblick, N.; McDarby, P.; McDonnell, D.; McDonald, B.; Karkun, R. Intelligent biofeedback using an immersive competitive environment. In Online Proceedings for the Designing Ubiquitous Computing Games Workshop, Ubicomp 2001, Atlanta GA, September 2001.
2. Hudlicka, E. Affective computing for game design. In Proceedings of the 4th International North American Conference on Intelligent Games and Simulation, Montreal, QB, Canada, 13–15 August 2008; pp. 5–12.
3. Tijs, T.; Brokken, D.; Ijsselsteijn, W. Creating an emotionally adaptive game. In *International Conference on Entertainment Computing*; Springer: Berlin, Heidelberg, Germany, 2008; Volume 5309, pp. 122–133.
4. Poria, S.; Cambria, E.; Bajpai, R.; Hussain, A. A review of affective computing: From unimodal analysis to multimodal fusion. *Inf. Fusion* **2017**, *37*, 98–125.
5. Gomes, P.V.; Marques, A.; Pereira, J.; Correia, A.; Donga, J.; Sá, V.J. E-emotion capsule: As artes digitais na criação de emoções. In Proceedings of the 9th International Conference on Digital and Interactive Arts (ARTECH 2019). Association for Computing Machinery, New York, NY, USA; Article 86, 1–4. DOI: https://doi.org/10.1145/3359852.3359962.

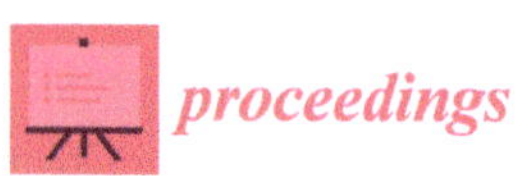

Proceedings

CultUnity3D: A Virtual Spatial Ecosystem for Digital Engagement with Cultural Heritage Sites †

Estefanía López Salas [1,*], Adrián Xuíz García [2], Ángel Gómez [3] and Carlos Dafonte [3]

[1] School of Architecture, Department of Project Design, Urbanism and Composition, Universidade da Coruña, Campus de A Zapateira s/n, 15071 A Coruña, Spain
[2] CITIC, Universidade da Coruña, Campus de Elviña s/n, 15071 A Coruña, Spain; adrian.xuizg@udc.es
[3] CITIC—Department of Computer Science and IT, Universidade da Coruña, Campus de Elviña s/n, 15071 A Coruña, Spain; angel.gomez@udc.es (Á.G.); carlos.dafonte@udc.es (C.D.)
* Correspondence: estefania.lsalas@udc.es

† Presented at the 3rd XoveTIC Conference, A Coruña, Spain, 8–9 October 2020.

Published: 20 August 2020

Abstract: In order to help enhance public outreach and understanding of historical sites, we developed a virtual spatial ecosystem called CultUnity3D. It consists of a set of components specifically implemented within the Unity engine that enable the user to virtually explore spatial changes over time in two different modes, and to learn about the past of a built environment through the integration of and interaction with research sources and narrative. Although we built CultUnity3D for a particular case study, which is the monastic site of San Julián de Samos (Spain), this in-progress virtual ecosystem has been thought out and designed for continued and reusable development.

Keywords: virtual reality; 3D modeling; Unity; spatiotemporal simulation; digital art and architectural history

---

1. Introduction

Digital representations have become essential tools to increase public awareness and enhance understanding of cultural heritage sites. The ICOMOS charter (2008) highlights the importance of an effective interpretation and presentation of places of historical and cultural significance as a way to communicate their values, to ensure their conservation, and to promote further "interest, learning, experience, and exploration" by the public at large ([1], p. 4).

However, to digitally represent and effectively communicate the research about historical built environments is a challenge. New complexities arise when we aim to create a virtual product to disseminate knowledge about historical sites that is not only simple and attractive to engage broader audiences, but it is also as intellectually and technically rigorous as other longer established methods of dissemination. For this purpose, the London Chapter for the Computer-Based Visualization of Cultural Heritage (2009) expresses both the need to clearly document the data sources (physical remains, archival documents, photographs, etc.) on which the digital re-constructions are based and the status of the knowledge that they represent (evidence vs. hypothesis) within an integral final output [2] (pp. 2, 7–8). In this paper, we report our in-process work to create CultUnity3D, which is the first release of a virtual ecosystem that aims to make the spatial transformations of the monastic site of San Julián de Samos over time more accessible, meaningful and experiential to all-aged audiences.

2. Project Background

San Julián de Samos is the most important Benedictine monastery in the French Way to Santiago de Compostela in Galicia. It has been listed as a historical monument since 1944 as it presents significant values from the past. Founded around the mid-sixth century, this building and its context evolved for centuries through successive constructions, re-constructions and demolitions that created, changed, or caused the loss of the monastic site at Samos.

Based on previous studies, archival documentation, and on-site investigation, we generated a series of phased 2D maps and 3D models that address and visualize questions of change over time [3]. As researchers, these computer-aided design (CAD) outputs help us to gain knowledge about the site, but the following questions remain: How can we communicate this outside academia and, by doing so, promote learning about cultural heritage? In which way could we represent the makings and shaping of a historical site as an ongoing process where both space and time are involved? Is it possible to create a digital ecosystem where multiple architectural models are visualized together with disparate historical datasets opened to scientific assessment and public engagement in an accessible and interactive way? To give a proper answer to the previous questions, we developed CultUnity3D.

3. Functionalities and Components

For the development of this virtual ecosystem, the designs in CAD format of the real environment to be modeled have been used as a base, converting and adjusting them to a 3D format for subsequent import into the Unity engine. The creation of the customizable functionalities and components (controllers, buttons, effects, etc.) have been implemented using C# scripts integrated inside Unity [4]. The functionalities and components of CultUnity3D are as follows.

3.1. Space and Time Interaction

The movement of the user through the virtual model utilizes existing Unity navigation tools. In this way, it is able to experience movement in and around the building (spatial interaction). However, to discover how the monastic site was transformed at distinct time periods (time interaction), we implement a specific component: AgeController.

3.2. Mode Controller: First Person Control and Guided Time Travel

Two types of Virtual Reality exploration are available through the ModeController script: First Person Control, which enables the user to choose the view and path to be explored at will ([5], pp. 147–148), and Guided Time Travel, which turns the user into a passive observer to watch a predefined tour set.

3.3. Integration of Research Sources and Narrative

The virtual reconstruction of the historical site in CultUnity3D aims to be a self-explanatory product where we integrate the architectural models and plans, the research sources and a brief narrative of the site's biography [6], pp. 488–490. To allow users access to data connected to what they are exploring in the scene, thus far, we have defined the components AnimationButton, FocusButton, ImageAndAudioButton and SwitchButton (see Figure 1).

(a) (b)

Figure 1. Screenshots of two interactive buttons designed: (**a**) to display graphic research sources, and (**b**) to give access to a brief textual and audio narrative.

3.4. Reusable and User-friendly Infrastructure

Instead of building a project-specific solution, CultUnity3D aims to provide a tool set of reusable features for future case studies, where the visualization of spatial change over time is the main art historical question to tackle and communicate. Moreover, our in-progress ecosystem is conceived to enable a user-friendly future development and customization run not only by experts in programming, but also scholars in art, architectural and urban history. CultUnity3D can also be easily adapted to different display devices, including touch screens or 3D glasses for virtual reality environments.

4. Conclusions

Through CultUnity3D, we create a virtual ecosystem where the understanding of a cultural heritage site with a complex long life is possible. This is not a completed project, but a first step with open opportunities, and also new challenges to face in future developments. For instance, we must work on the usability of the navigation controls and the user interface (menu, instructions, graphical UI elements, etc.). We also need to implement new components to increase the interaction between the user and the model in different devices. The spatiotemporal functionality needs to be improved (time slider, multi-mode viewer, …) to provide a different understanding than is possible with static images. We need to work on the Guided Time Travel to offer the user thoughtful, predefined tours within a virtual historical site. In this sense, we also consider it important to continue working on the integration of historical data and its display in the user interface. The visualization of spatial changes over time with disparate research sources and physical evidence for public dissemination demands new answers, but it also opens up new ways to think about and produce knowledge.

Author Contributions: All authors have equally contributed to this paper. All authors have read and agreed to the published version of the manuscript.

Funding: This research received no external funding.

Conflicts of Interest: The authors declare no conflict of interest.

References

1. *The ICOMOS Charter for the Interpretation and Presentation of Cultural Heritage Sites*; ICOMOS: Paris, France, 2008. Available online: https://www.icomos.org/en/resources/charters-and-texts (accessed on 13 June 2020).
2. *The London Chapter for the Computer-Based Visualization of Cultural Heritage*; London, UK, 2009. Available online: http://www.londoncharter.org/downloads.html (accessed on 13 June 2020).
3. López Salas, E. San Julián de Samos-Lugo, Estudio e Interpretación del Diseño Monástico y su Evolución. Ph.D. Thesis, Universidade da Coruña, A Coruña, Spain, 2015.
4. Xuíz García, A. Sistema de Representación Virtual Interactiva do Modelo Dunha Estrutura Arquitectónica de alto valor Cultural. Ph.D. Thesis, Universidade da Coruña, A Coruña, Spain, 2020.
5. Tan, B.K; Hafizur, R. Virtual Heritage: Reality and Criticism. In *Joining Languages, Cultures and Visions: CAADFutures 2009*; Tidalfi, T., Dorta, T., Eds.; Presses de l'Université de Montréal: Montréal, QC, Canada, 2009, pp. 143–156.

6. Wendell, A.; Ozludil Altin, B.; Thompson, U. Prototyping a Temporospatial Simulation Framework: Case of an Ottoman Insane Asylum. In *Complexity & Simplicity-Proceedings of the 34th International Conference on Education and Research in Computer Aided Architectural Design in Europe*; Herneoja, A., Österlund, T., Markkanen, P., Eds.; eCAADe and Oulu School of Architecture: Brussels, Belgium, 2016; Volume 2, pp. 485-491.

Proceedings

Web Server and R Library for the Calculation of Markov Chains Molecular Descriptors †

Paula Carracedo-Reboredo [1], Cristian R. Munteanu [1,2], Humbert González-Díaz [3,4] and Carlos Fernandez-Lozano [1,2,*]

1 Departament of Computer Science and Information Technologies, Faculty of Computer Science, Universidade da Coruña, Campus Elviña s/n, 15071 A Coruña, Spain; paula.carracedo@udc.es (P.C.-R.); c.munteanu@udc.es (C.R.M.)
2 CITIC-Research Center of Information and Communication Technologies, Universidade da Coruña, 15071 A Coruña, Spain
3 Basque Center for Biophysics, University of the Basque Country UPV/EHU, 48940 Leioa, Bilbao, Spain; humberto.gonzalezdiaz@ehu.es
4 IKERBASQUE, Basque Foundation for Science, 48011 Bilbao, Spain
* Correspondence: carlos.fernandez@udc.es

† Presented at the 3rd XoveTIC Conference, A Coruña, Spain, 8–9 October 2020.

Published: 20 August 2020

Abstract: Markov Chain Molecular Descriptors (MCDs) have been largely used to solve Cheminformatics problems. The software to perform the calculation is not always available for general users. In this work, we developed the first library in R for the calculation of MCDs and we also report the first public web server for the calculation of MCDs online that include the calculation of a new class of MCDs called Markov Singular values. We also report the first Cheminformatics study of the biological activity of 5644 compounds against colorectal cancer.

Keywords: Markov Chains; online tool; R; colorectal cancer

---

1. Introduction

Cheminformatics models are able to predict different outputs in complex molecular systems. On the other hand, colorectal cancer (CRC) is the third most commonly occurring cancer in men and the second in women, having a mortality of approximately 56% of the patients [1]. Although a number of compounds for anti-CRC activity have been synthetized and tested, the possibility of coming across an effective drug is still too low [2]. Markov Chain Molecular Descriptors (MCDs) have been largely used to solve Cheminformatics problems, and the calculation is done very often using specific software not ever available for general users. In this work, we developed the first library in R [3] for the calculation of MCDs and the first public web server for the calculation online that includes the calculation of a new class of MCDs called Markov Singular indices. We report a case study; we illustrated the use of those molecular descriptors in the study of active compounds against colorectal cancer (CRC).

2. Materials and Methods

We proposed an implementation in R of the algorithm for calculation of MCDs that can calculate two drug topological indices (TIs) families: Markov Mean Properties (MMPs) and Markov Singular Values of Transition Probabilities (MMSVs). The combination of the RMarkov.mol with the RRegrs package generates a powerful and fast R tool for designing QSAR (Quantitative structure-activity relationship) regression models. We obtained 5644 preclinical assays of CRC active compounds from ChEMBL and calculated the MCDs using our web server to simplify the process.

3. Results

Figure 1 shows the user interface of Markov Chemical Descriptors Calculator (MCDCalc) web server. This allows the calculation of molecular descriptors for each atomic property and type of atom. Smiles formulas can be read from the text file or can be individually pasted on screen textbox.

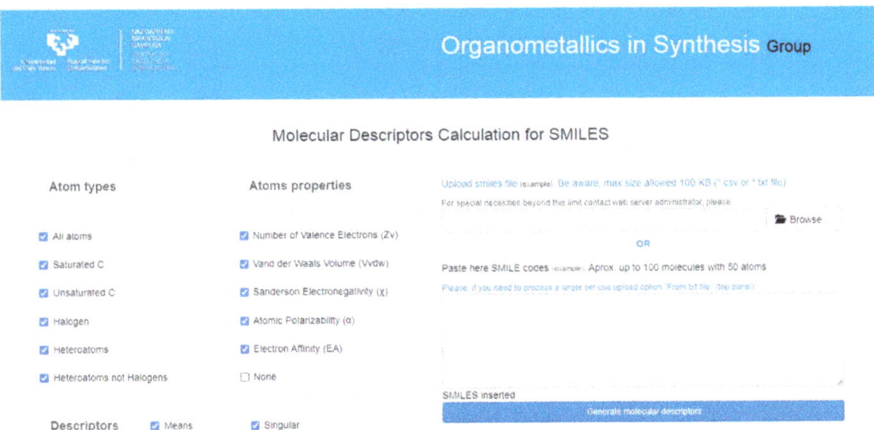

Figure 1. Online web server.

We also used the molecular descriptors as input for the RRegrs [4] package in order to find better regression models. Random Forest (RF), Support Vector Machines (SVM), Neural Networks (NN), and Partial Least Squares (PLS) regression methods have been tested and the results are presented in Table 1.

Table 1. Results for RF, SVM, NN and PLS.

Method	Training		Test	
	R^2	RMSE	R^2	RMSE
RF	0.907	0.101	0.926	0.093
SVM	0.868	0.122	0.866	0.128
NN	0.849	0.132	0.829	0.143
PLS	0.801	0.155	0.775	0.167

4. Conclusions

We have developed the first library in R for the calculation of MCDs, and the first public web server for the calculation of MCDs online that includes the calculation of Markov Singular values which are useful to predict the activity prediction of anti-colorectal cancer compounds. The RF regression model showed the best results.

References

1. Riihimaki, M.; Hemminki, A.; Sundquist, J.; Hemminki, K. Pat-terns of Metastasis in Colon and Rectal Cancer. *Sci. Rep.* **2016**, *6*, 1–9, doi:10.1038/srep29765.
2. Jhanwar, B.; Sharma, V.; Singla, R.K.; Shrivastava, B. QSAR-Hansch Analysis and Related Approaches in Drug Design. *Phar.-Macol. Online Newsl.* **2011**, *1*, 306–344.

3. Santiago, C.B.; Guo, J.Y.; Sigman, M.S. Predictive and Mechanistic Multivariate Linear Regression Models for Reaction Development. *Chem. Sci.* **2018**, *9*, 2398–2412, doi:10.1039/c7sc04679k.
4. Tsiliki, G.; Munteanu, C.R.; Seoane, J.A.; Fernandez-Lozano, C.; Sarimveis, H.; Willighagen, E.L. RRegrs: An R package for computer-aided model selection with multiple regression models. *J. Cheminform.* **2015**, *7*, 46, doi:10.1186/s13321-015-0094-2.

Proceedings

Validation of Self-Quantification Xiaomi Band in a Clinical Sleep Unit †

Francisco José Martínez-Martínez [1], Patricia Concheiro-Moscoso [1,*], María Del Carmen Miranda-Duro [1], Francisco Docampo Boedo [2], Francisco Javier Mejuto Muiño [2] and Betania Groba [1]

1 CITIC, TALIONIS Group, Elviña Campus, Universidade da Coruña (University of A Coruña), 15071 A Coruña, Spain; f.martinezm@udc.es (F.J.M.-M.); carmen.miranda@udc.es (M.D.C.M.-D.); b.groba@udc.es (B.G.)

2 Hospital San Rafael, Las Jubias, 15009 A Coruña, Spain; fdocampo@imqsanrafael.es (F.D.B.); fmejmui@gmail.com (F.J.M.M.)

* Correspondence: patricia.concheiro@udc.es

† Presented at the 3rd XoveTIC Conference, A Coruña, Spain, 8–9 October 2020.

Published: 21 August 2020

Abstract: Polysomnography (PSG) is currently the accepted gold standard for sleep studies, as it measures multiple variables that lead to a clear diagnosis of any sleep disorder. However, it has some clear drawbacks, since it can only be performed by qualified technicians, has a high cost and complexity and is very invasive. In the last years, actigraphy has been used along PSG for sleep studies. In this study, we intend to assess the capability of the new Xiaomi Mi Smart Band 5 to be used as an actigraphy tool. Sleep measures from PSG and Xiaomi Mi Smart Band 5 recorded in the same night will be obtained and further analysed to assess their concordance. For this analysis, we perform a paired sample t-test to compare the different measures, Bland–Altman plots to evaluate the level of agreement between the Mi Band and PSG and Epoch by Epoch analysis to study the ability of the Mi Band to correctly identify PSG-defined sleep stages. This study belongs to the research field known as participatory health, which aims to offer an innovative healthcare model driven by the patients themselves, leading to civic empowerment and self-management of health.

Keywords: sleep; polysomnography; participatory health; Xiaomi Mi Smart Band 5; Internet of Things

1. Introduction

Sleep has considerable implications in our daily life, and it is crucial to effectively accomplish basic vital functions. People whose sleep quality is poor exhibit different sleep disturbances, such as the fragmentation of the different sleep stages, night arousals and a greater will of having diurnal naps. Moreover, in the most serious cases, sleep disorders like insomnia, hypersomnia or sleep apnoea may also appear [1]. Sleep Units are specialised areas for diagnosis and treatment of all these sleep disorders and offer a wide range of diagnostic tests (e.g., PSG or Multiple Latency Test) [2]. Despite providing multiple possibilities, PSG is considered as the most reliable instrument for the measure of sleep parameters by the scientific community. However, PSG also implies high invasiveness and cost, as well as specialised technicians to use it. Hence, different approaches for sleep assessment using wearable devices are being compared with PSG in order to achieve their maximum possible quality for sleep studies, so they can be used as a complementary tool [3]. In this study, the validation of the sleep data recorded by Xiaomi wristbands is assessed to determine whether these measurements are reliable enough for people to evaluate their own sleep. On this matter, previously conducted studies about these type of wearables have shown a high accuracy and sensitivity, a low specificity and a poorly significant and limited estimation of sleep/wakefulness states [4].

2. Methods

2.1. Design of the Study

This is an observational, analytic, longitudinal, pilot study whose aim is to demonstrate that data collecting instruments, along with their management, are viable and effective. Different variables from the population of interest will be observed and recorded without any direct intervention, so as to establish causality associations between these variables. It is considered as longitudinal, since the tracking of the variables will be performed during six months, continually (and occasionally) recording and monitoring sleep quality. A difference of >15 min in deep sleep measures between Xiaomi and PSG will be considered as clinically relevant in this study. Accepting a 0.05 α risk and a 0.1 β risk (statistical power of 90%) in a bilateral contrast, a population of 43 patients is needed to detect a difference that is $\geq$15 min. According to previous studies, a $\pm$30 SD will be assumed. Only patients that perform a medical test at the Sleep Unit from San Rafael Hospital and are >18 years old will be asked to participate. Legal and ethical aspects that guarantee good clinical practice will be followed in this study. Therefore, the informed consent process will be carried out with all participants.

2.2. Data Collection and Analysis

In this study, PSG data in EDF+ format, the most accepted standard to exchange EEG and PSG data [5], will be obtained from patients that undergo a sleep study at the Sleep Unit of San Rafael Hospital. In addition, patients will be given a Xiaomi Mi Smart Band 5 that will measure their sleep along with PSG for one night. By doing so, we will be able to compare both recordings in order to assess if these wearable devices show results in concordance with PSG. Besides, software to obtain data second by second from Xiaomi bands was developed by the TALIONIS Group, since these wearables export daily data by default, which hinder the analysis.

2.2.1. Variables of Interest

Table 1 shows the features that will be extracted from our data. Numeric variables will be shown as mean (M) and standard deviation (SD), including their range, minimums and maximums.

Table 1. Summary of the features of interest for our study

Variable	Description	Dimension
Time in Bed (TIB)	Total time the patient is laying down	min
Sleep Onset Latency (SOL)	Length of time from full wakefulness to sleep	min
Wake After Sleep Onset (WASO)	Periods of wakefulness after defined sleep onset	min
Sleep Efficiency (SE)	Time spent asleep/Time in Bed $*$ 100	%
Light sleep	$N1 + N2$ sleep stages	min
Deep sleep	$N3$ sleep stage	min
REM sleep	-	min

2.2.2. Statistical Analysis

After preprocessing the data (described in Figure 1), and following the analysis performed in the available literature for these validation studies [6], the summary of the aforementioned variables for both the Xiaomi Mi Smart Band 5 and PSG will be compared by using a paired sample t-test to study if there are significant differences between the means of each Xiaomi-PSG variable. To evaluate the level of agreement between Xiaomi and equivalent PSG sleep measures, we will use Bland–Altman plots. Since we are interested in the ability of Xiaomi devices to correctly identify sleep stages, Epoch by Epoch (EBE) analysis will be used to calculate the sensitivity (proportion of epochs identified as sleep by PSG that are correctly classified by the device), specificity (proportion of epochs identified as awake by PSG that are correctly classified by the device), agreement between both PSG and Xiaomi device in light sleep (proportion of PSG F1 + F2 epochs identified as light sleep by the device), deep sleep (proportion of PSG F3 + F4 epochs identified as deep sleep by the device) and REM sleep (proportion of PSG REM identified as REM by the device) identification.

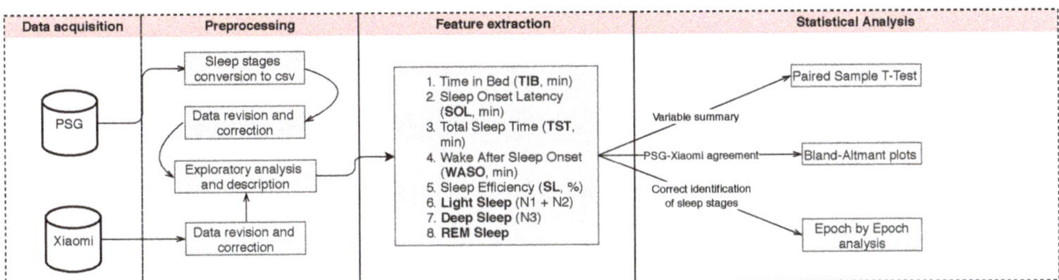

Figure 1. Sleep data validation workflow.

3. Conclusions

This study offers a promising approach to assess whether wearable devices, in our case the Xiaomi Mi Smart Band 5, are able to correctly record our sleep. Even though these devices are not expected to replace polysomnography studies, they may be used as an initial evaluation for users to manage their own sleep quality and, if necessary, visit their doctor or simply change some habits.

Author Contributions: Conceptualization, F.J.M.-M., P.C.-M., and B.G.; methodology, F.J.M.-M., P.C.-M., and B.G.; investigation, F.J.M.-M., P.C.-M., and B.G.; writing–original draft preparation, F.J.M.-M., P.C.-M., and M.d.c.M.-D.; writing–review and editing, B.G., F.J.M.M. and F.D.B.; visualization,F.J.M.-M., P.C.-M., and M.d.c.M.-D.; supervision, B.G., F.J.M.M. and F.D.B.; project administration, B.G., F.J.M.M. and F.D.B.; funding acquisition, P.C.-M. and M.d.c.M.-D. All authors have read and agreed to the published version of the manuscript.

Acknowledgments: The authors disclose the receipt of the following financial support for the research, authorship and/or publication of this article: All the economic costs involved in the study will be borne by the research team. This work is supported in part by some grants from the European Social Fund 2014-2020. CITIC (Research Centre of the Galician University System) and the Galician University System (SUG) obtained funds through Regional Development Fund (ERDF) with 80%, Operational Programme ERDF Galicia 2014-2020 and the remaining 20% by the Secretaría Xeral de Universidades of the Galician University System (SUG). Specifically, the author P.C.M obtained a scholarship (Ref. ED481A-2019/069) and the author M.d.c.M.-D. (Ref. ED481A 2018/205) to develop the PhD thesis. Furthermore, the diffusion and publication of this research is funded by the CITIC, Research Centre of the Galician University System with the support previously mentioned (Ref ED431G 2019/01).

Conflicts of Interest: The authors declare no conflict of interest.

References

1. Chen, J.H.; Waite, L.; Kurina, L.M.; Thisted, R.A.; McClintock, M.; Lauderdale, D.S. Insomnia symptoms and actigraph-estimated sleep characteristics in a nationally representative sample of older adults. *J. Gerontol.* **2015**, *70*, 185–192. doi:10.1093/gerona/glu144.
2. Rundo, J.V.; Downey, R. Polysomnography. *Handb. Clin. Neurol.* **2019**, *160*, 381–392. doi:10.1016/B978-0-444-64032-1.00025-4.
3. Shelgikar, A.V.; Anderson, P.F.; Stephens, M.R. Sleep Tracking, Wearable Technology, and Opportunities for Research and Clinical Care. *Chest* **2016**, *150*, 732–743. doi:10.1016/j.chest.2016.04.016.
4. Kahawage, P.; Jumabhoy, R.; Hamill, K.; de Zambotti, M.; Drummond, S.P. Validity, potential clinical utility, and comparison of consumer and research-grade activity trackers in Insomnia Disorder I: In-lab validation against polysomnography. *J. Sleep Res.* **2020**, *29*, 1–11. doi:10.1111/jsr.12931.

5. Kemp, B.; Olivan, J. European data format 'plus' (EDF+), an EDF alike standard format for the exchange of physiological data. *Clin. Neurophysiol.* **2003**, *114*, 1755–1761. doi:10.1016/S1388-2457(03)00123-8.
6. de Zambotti, M.; Rosas, L.; Colrain, I.M.; Baker, F.C. The Sleep of the Ring: Comparison of the ŌURA Sleep Tracker Against Polysomnography. *Behav. Sleep Med.* **2019**, *17*, 124–136. doi:10.1080/15402002.2017.1300587.

Proceedings

Designing an Open Source Virtual Assistant †

Anxo Pérez *, **Paula Lopez-Otero** and **Javier Parapar**

Information Retrieval Lab, Centro de Investigación en Tecnoloxías da Información e as Comunicacións (CITIC), Universidade da Coruña, 15071 A Coruña, Spain; paula.lopez.otero@udc.es (P.L.-O.); javier.parapar@udc.es (J.P.)

* Correspondence: anxo.pvila@udc.es; Tel.: +34-881-01-1276

† Presented at the 3rd XoveTIC Conference, A Coruña, Spain, 8–9 October 2020.

Published: 21 August 2020

Abstract: A chatbot is a type of agent that allows people to interact with an information repository using natural language. Nowadays, chatbots have been incorporated in the form of conversational assistants on the most important mobile and desktop platforms. In this article, we present our design of an assistant developed with open-source and widely used components. Our proposal covers the process end-to-end, from information gathering and processing to visual and speech-based interaction. We have deployed a proof of concept over the website of our Computer Science Faculty.

Keywords: conversational assistant; chatbot; question answering; natural language processing; crawling; information retrieval

---

1. Introduction

Nowadays, conversational systems are part of our daily routines [1]. Tech giants are aware of their relevance, and they are incorporating these assistants to their platforms. Microsoft Cortana or Apple Siri are popularly used examples. Some companies, such as Google with the Assistant or Amazon with Alexa, are even manufacturing dedicated devices. Moreover, these services offer great opportunities for customer support [2]. Many companies' websites are gradually enabling conversational capacities to help users discover products and information [3,4]. On the other hand, voice commands have gained much attraction for user interaction [5]. There is a critical tendency to move towards audio controls over tactile interfaces [6]. The inclusion of voice capabilities was, therefore, a straightforward improvement for conversational agents. Apart from the business value of the technology, voice-enabled assistants are truly useful for people with functional diversity [7].

In this article, we propose an architecture for a conversational assistant. As we mentioned, our proposal covers the process end-to-end. We apply information retrieval, natural language processing, machine learning, and speech technologies to cover data acquisition to audio response and user questions.

2. Proposal

As mentioned previously, our architectural design involves all stages of the process. The system covers everything from information gathering and processing to visual- and speech-based interaction. For that, we have used models and techniques from different information processing fields. In this section, we will explain the process we have followed and the description of the technologies used in the development.

A web-crawler is in charge of the first step of the information-processing pipeline. In this phase, we retrieved the information from the domain webpages, and kept those documents up-to-date. For this task, we used Scrapy (https://scrapy.org/), a popular web scraper. It saves the data from the Internet and creates a repository containing the files to be indexed. The website (https://www.fic.udc.es/) contains documents both in Spanish and Galician. The second phase corresponds to text-processing, which includes sentence-splitting and indexing. We used ElasticSearch (https://www.elastic.co/), a distributed search engine based on Lucene, for both indexing and searching. We propose the use of the ElasticSearch identification component for tagging the documents. As we were building a conversational system, we indexed the data at both the document- and sentence-level. Indexing isolated sentences allowed us to answer many of the user's questions concisely and directly.

Our design accepts the user's input, both through writing and spoken queries. For processing voice queries, we used Kaldi (https://kaldi-asr.org/) for Automatic Speech Recognition (ASR). Finally, we provided a system response in text and audio format using Cotovia (http://gtm.uvigo.es/cotovia). Here, we again used language identification models to process user inputs and outputs correctly. In the case of voice interaction, we trained the automatic speech recognition language models with specific domain lexica [8]. On the voice response side, the system reproduces the responses, selecting the language accordingly to the user input. In this case, we used Cotovia pre-trained models to perform speech synthesis [9].

One crucial problem of spoken document retrieval is term misrecognition. This problem provokes the inability to process the information need correctly. ASR misrecognition produces term mismatch between user input and document content. We used efficient state-of-the-art retrieval models [10] based on n-gram decomposition in dealing with it. The system processes both searchable content and user input in that way to allow fast and robust query matching. These models achieve state-of-the-art effectiveness figures, while also being quite efficient [11].

For answering information needs, we designed a four-level cascade system. First, the system tries to classify the user's intent in some predefined structured tasks (e.g., the timetable for a subject or the date of an exam). If the input falls onto one of those categories, the answer is processed according to the defined pipelines. Second, if that was not the case, the system attempts to provide a direct answer to the specific user question. For that, we propose to use two approaches: best-sentence-matching and BERT-based question-answering [12]. Thirdly, there is no satisfactory direct answer, the system tries to provide the best document answer. Finally, if the system does not rank satisfory documents, it asks the user to reformulate the question.

Architecturally speaking, we are thus using a basic client-server application. The web client communicates with the backed-through rest services and WebSocket APIs . For the web interface, we used BotUI (https://github.com/botui/botui), a very intuitive Javascript library for conversational interfaces. The server contains the implementation of the different REST endpoints and WebSocket APIS for processing audio streams.

3. Conclusions and Future Work

In this article, we presented our design of a conversational assistant based on open-source and well-known technologies. Even though we exemplified the design for a specific web domain for its implementation, the architecture introduced here can consume other information repositories, such as different enterprises data information or any databases.

There are many avenues for future work. We propose to improve the architecture with advanced Natural Language Generation (NLG) capacities. The fine-tuning of acoustic models for specific language variants is another interesting research address. When not depending on languages, such as Galician, we would favor the use of Tacotron as the TTS engine [13].

Funding: This work was supported by projects RTI2018-093336-B-C22 (MCIU & ERDF) and GPC ED431B 2019/03 (Xunta de Galicia & ERDF). Also, this work has received financial support from CITIC, *Centro de Investigación del Sistema universitario de Galicia*, which is financial supported by *Consellería de Educación, Universidade e Formación Profesional* of the *Xunta de Galicia* through the ERDF (80%) and *Secretaría Xeral de Universidades* (20%), (Ref ED431G 2019/01).

Conflicts of Interest: The authors declare no conflict of interest.

References

1. Masche, J.; Le, N.T. A Review of Technologies for Conversational Systems. In *International Conference on Computer Science, Applied Mathematics and Applications*; Springer: Cham, Switzerland, 2018; pp. 212–225, doi:10.1007/978-3-319-61911-8_19.
2. Brandtzaeg, P.B.; Følstad, A. Why People Use Chatbots. In *Internet Science*; Kompatsiaris, I., Cave, J., Satsiou, A., Carle, G., Passani, A., Kontopoulos, E., Diplaris, S., McMillan, D., Eds.; Springer International Publishing: Cham, Switzerland, 2017; pp. 377–392.
3. Chung, M.; Ko, E.; Joung, H.; Kim, S.J. Chatbot e-service and customer satisfaction regarding luxury brands. *J. Bus. Res.* **2018**. doi:10.1016/j.jbusres.2018.10.004.
4. Majumder, A.; Pande, A.; Vonteru, K.; Gangwar, A.; Maji, S.; Bhatia, P.; Goyal, P. An Approach Based on Category-Sensitive Retrieval. In *European Conference on Information Retrieval Automated Assistance in E-commerce*; Springer: Cham, Switzerland, 2018; pp. 604–610, doi:10.1007/978-3-319-76941-7_51.
5. Bhalla, A. An exploratory study understanding the appropriated use of voice-based Search and Assistants. In *IndiaHCI'18: Proceedings of the 9th Indian Conference on Human Computer Interaction*; Association for Computing Machinery: New York, NY, USA, 2018; pp. 90–94, doi:10.1145/3297121.3297136.
6. Porcheron, M.; Fischer, J.E.; Reeves, S.; Sharples, S. Voice Interfaces in Everyday Life. In Proceedings of the 2018 CHI Conference on Human Factors in Computing Systems, CHI '18, Montreal, QC, Canada, 21–26 April 2018; Association for Computing Machinery: New York, NY, USA, 2018; pp. 1–12, doi:10.1145/3173574.3174214.
7. Brewer, R.N.; Findlater, L.; Kaye, J.J.; Lasecki, W.; Munteanu, C.; Weber, A. Accessible Voice Interfaces. In Proceedings of the Companion of the 2018 ACM Conference on Computer Supported Cooperative Work and Social Computing, CSCW '18, New York, NY, USA, 3–7 November 2018; Association for Computing Machinery: New York, NY, USA, 2018; pp. 441–446, doi:10.1145/3272973.3273006.
8. Povey, D.; Ghoshal, A.; Boulianne, G.; Burget, L.; Glembek, O.; Goel, N.; Hannemann, M.; Motlíček, P.; Qian, Y.; Schwarz, P.; et al. The Kaldi speech recognition toolkit. In Proceedings of the IEEE 2011 Workshop on Automatic Speech Recognition and Understanding, Waikoloa, HI, USA, 11–15 December 2011.
9. Banga, E.R.; Mateo, C.G.; Pazó, F.J.M.; González, M.G.; Magariños, C. Cotovía: An open source TTS for Galician and Spanish. In Proceedings of the "IberSPEECH 2012:" "VII Jornadas en Tecnología del Habla" " and III Iberian SLTech Workshop", Madrid, Spain, 21–23 November 2012.
10. Lopez-Otero, P.; Parapar, J.; Barreiro, A. Statistical language models for query-by-example spoken document retrieval. *Multim. Tools Appl.* **2020**, *79*, 7927–7949, doi:10.1007/s11042-019-08522-z.
11. Lopez-Otero, P.; Parapar, J.; Barreiro, A. Efficient query-by-example spoken document retrieval combining phone multigram representation and dynamic time warping. *Inf. Process. Manag.* **2019**, *56*, 43–60. doi:10.1016/j.ipm.2018.09.002.
12. Carrino, C.P.; Costa-jussà, M.R.; Fonollosa, J.A.R. Automatic Spanish Translation of the SQuAD Dataset for Multilingual Question Answering. *arXiv* **2019**, arXiv:1912.05200.
13. Wang, Y.; Skerry-Ryan, R.; Stanton, D.; Wu, Y.; Weiss, R.J.; Jaitly, N.; Yang, Z.; Xiao, Y.; Chen, Z.; Bengio, S.; et al. Tacotron: Towards End-to-End Speech Synthesis. *arXiv* **2017**, arXiv:1703.10135.

Proceedings

Analysis of Separability of COVID-19 and Pneumonia in Chest X-ray Images by Means of Convolutional Neural Networks †

Joaquim de Moura [1,2,*], Lucía Ramos [1,2], Plácido L. Vidal [1,2], Jorge Novo [1,2] and Marcos Ortega [1,2]

1 Centro de investigación CITIC, Universidade da Coruña, 15071 A Coruña, Spain; l.ramos@udc.es (L.R.); placido.francisco.lizancos.vidal@udc.es (P.L.V.); jnovo@udc.es (J.N.); mortega@udc.es (M.O.)
2 Grupo VARPA, Instituto de Investigación Biomédica de A Coruña (INIBIC), Universidade da Coruña, 15006 A Coruña, Spain
* Correspondence: joaquim.demoura@udc.es; Tel.: +34-981-167-000 (ext. 1330)
† Presented at the 3rd XoveTIC Conference, A Coruña, Spain, 8–9 October 2020.

Published: 21 August 2020

Abstract: The new coronavirus (COVID-19) is a disease that is caused by severe acute respiratory syndrome coronavirus 2 (SARS-CoV-2). On 11 March 2020, the coronavirus outbreak has been labelled a global pandemic by the World Health Organization. In this context, chest X-ray imaging has become a remarkably powerful tool for the identification of patients with COVID-19 infections at an early stage when clinical symptoms may be unspecific or sparse. In this work, we propose a complete analysis of separability of COVID-19 and pneumonia in chest X-ray images by means of Convolutional Neural Networks. Satisfactory results were obtained that demonstrated the suitability of the proposed system, improving the efficiency of the medical screening process in the healthcare systems.

Keywords: computer-aided diagnosis; chest X-ray imaging; COVID-19; pneumonia; deep learning

1. Introduction

The coronavirus disease 2019 (COVID-19) disease caused by severe acute respiratory syndrome coronavirus 2 (SARS-CoV-2) has resulted in an unprecedented public health crisis. This highly infectious disease was first identified in the city of Wuhan in Hubei province, China in December 2019. The number of confirmed cases is more than 7.5 million, affecting more than 213 countries, including 287,399 deaths according to the World Health Organization reports. COVID-19 has rapidly progressed to become a global pandemic, causing an unprecedented impact on the health, social, and economic well-being of people around the world. One of the most effective ways of limiting this relevant pandemic disease is the early, rapid, and accurate diagnosis and treatment of infected patients. In this sense, chest X-ray images are the most common and widely available diagnostic imaging technology, playing a crucial role in clinical care and epidemiological studies of confirmed or suspected of COVID-19 cases.

2. Methodology

Thus, in this work, we present a novel fully automatic methodology for the analysis of separability of COVID-19 and pneumonia-infected lungs in chest X-ray images, given the high level of similarity between these two lung diseases. To this end, different complementary deep learning-based approaches that are based on a densely convolutional network architecture are adapted to better analyze the distinctive clinical patterns of these relevant diseases [1].

3. Results and Conclusions

We perform different experiments using two publicly available chest X-ray image datasets in order to validate the designed methodology [2,3]. Satisfactory results were obtained that demonstrated the suitability of the proposed methodology to facilitate early diagnosis and, thus, enable support to the clinical decision-making process in this pandemic scenario. Figure 1 show representative examples of graphical representations of heatmaps based on predictions of pathological regions in the chest X-ray images.

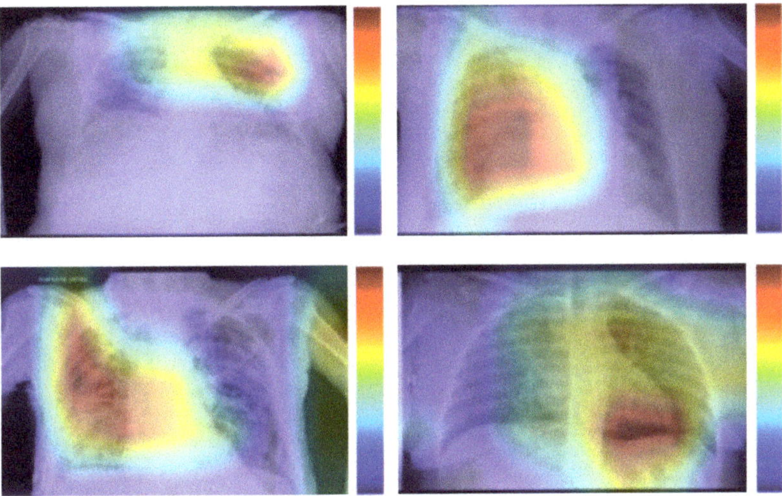

Figure 1. Results of the proposed methodology for the analysis of separability of COVID-19 and pneumonia-infected lungs in chest X-ray images. 1st row, chest X-ray images from patients infected with pneumonia. 2nd row, chest X-ray images from patients infected with COVID-19.

Author Contributions: J.d.M., L.R. and P.L.V. contributed to the analysis and design of the computer methods and the experimental evaluation methods. J.d.M., J.N. and M.O. contributed with domain-specific knowledge. All the authors performed the result analysis. J.d.M. was in charge of writing the manuscript, and all the authors participated in its critical revision and final approval. All authors have read and agreed to the published version of the manuscript.

Funding: This work is supported by the Instituto de Salud Carlos III, Government of Spain and FEDER funds of the European Union through the DTS18/00136 research projects and by the Ministerio de Ciencia, Innovación y Universidades, Government of Spain through the RTI2018-095894-B-I00 research projects, as well as through Ayudas para la formación de profesorado universitario (FPU), Ref. FPU18/02271. Also, this work has received financial support from the European Union (European Regional Development Fund—ERDF) and the Xunta de Galicia, Centro de Investigación del Sistema Universitario de Galicia, Ref. ED431G 2019/01.

Conflicts of Interest: The authors declare no conflict of interest.

References

1. De Moura, J.; Ramos, L.; Vidal, P.L.; Cruz, M.; Abelairas, L.; Castro, E.; Ortega, M. Fully automatic deep convolutional approaches for the analysis of covid-19 using chest X-ray images. *medRxiv* **2020**, 1–13.

2. Pneumonia Detection Challenge, Radiological Society of North America (RSNA). Available online: https://www.kaggle.com/c/rsna-pneumonia-detection-challenge/ (accessed on 10 April 2020)
3. COVID-19 DATABASE, Italian Society of Medical Radiology (SIRM). Available online: https://www.sirm.org/category/senza-categoria/covid-19/ (accessed on 10 April 2020)

Proceedings

Fully Automatic Retinal Vascular Tortuosity Assessment Integrating Domain-Related Information †

Lucía Ramos [1,2,*], Jorge Novo [1,2], José Rouco [1,2], Stéphanie Romeo [3], María D. Álvarez [3] and Marcos Ortega [1,2]

1 Centro de Investigación CITIC, Universidade da Coruña, 15071 A Coruña, Spain; j.novo@udc.es (J.N.); jose.rouco@udc.es (J.R.); m.ortega@udc.es (M.O.)
2 Grupo VARPA, Instituto de Investigación Biomédica de A Coruña (INIBIC), Universidade da Coruña, 15006 A Coruña, Spain
3 Servizo de Oftalmoloxía, Complexo Hospitalario Universitario de Ferrol, Ferrol, 15045 A Coruña, Spain; Stephanie.Romeo.Villadoniga@sergas.es (S.R.); maria.dolores.alvarez.diaz@sergas.es (M.D.Á.)
* Correspondence: l.ramos@udc.es; Tel.: +34-981167000 (ext. 5516)
† Presented at the 3rd XoveTIC Conference, A Coruña, Spain, 8–9 October 2020.

Published: 21 August 2020

Abstract: The fundus of the eye is the only part of the human body that allows a direct non-invasive observation of the circulatory system. Retinal vascular tortuosity presents a valuable potential for diagnostic and treatment purposes of relevant vascular and systemic diseases. This work presents a computational metric for the tortuosity characterization that combines mathematical representations of the vessel segments with anatomical properties of the fundus image such as the vessel caliber, the distance to the optic disc, the distance to the fovea and the distinction between arteries and veins. The evaluation of the prognostic performance shows that the incorporation of the domain-related information allows a reliable characterization of the retinal vascular tortuosity that provides a better representation of the expert perception.

Keywords: retinal circulation; vascular tortuosity; fundus images; computer-aided diagnosis; image analysis; clinical knowledge

1. Introduction

The retinal vascular tortuosity, characterized by an abnormal course of the blood vessels, has relevant potential as a clinical biomarker of a significative number of relevant diseases such as diabetic retinopathy, cerebrovascular disease, stroke, and ischemic heart disease, among others. Several computational approaches to address the assessment of the retinal vascular tortuosity have been proposed in the literature. These metrics are mostly based on mathematical properties of the vessel centerlines to describe the tortuosity according to the curvature, the amplitude, or the number of turns and twists of the vessel segments. However, the specialists, on the basis of their experience, besides the analysis of the vessel course, also consider additional domain-related parameters that are not incorporated in the computational metrics of reference. In particular, a set of anatomical factors, including the vessel caliber, the distinction between arteries and veins, the distance to the fovea, and the distance to the optic disc were identified as relevant for the tortuosity assessment. This work presents a computational methodology that combines mathematical properties of the retinal vessels with relevant domain-related information for a better representation of the expert criteria [1].

2. Methodology

The proposed computational metric consist of a tortuosity quantification based on a mathematical metric of reference [2] combined with the identified anatomical factors. For this purpose, first, the arterio-venous tree is extracted and decomposed into its constituent vessel segments. Then, a local tortuosity based on the mathematical representation of the vessel course weighted by the anatomical properties is computed for each individual vessel. The optimal configuration to incorporate the anatomical factors was set by means of a multi-objective optimization process. Finally, the local tortuosity values are combined in order to compute the global tortuosity measurement for the whole retina. Figure 1 shows examples of the intermediate steps for the computational tortuosity measuremet.

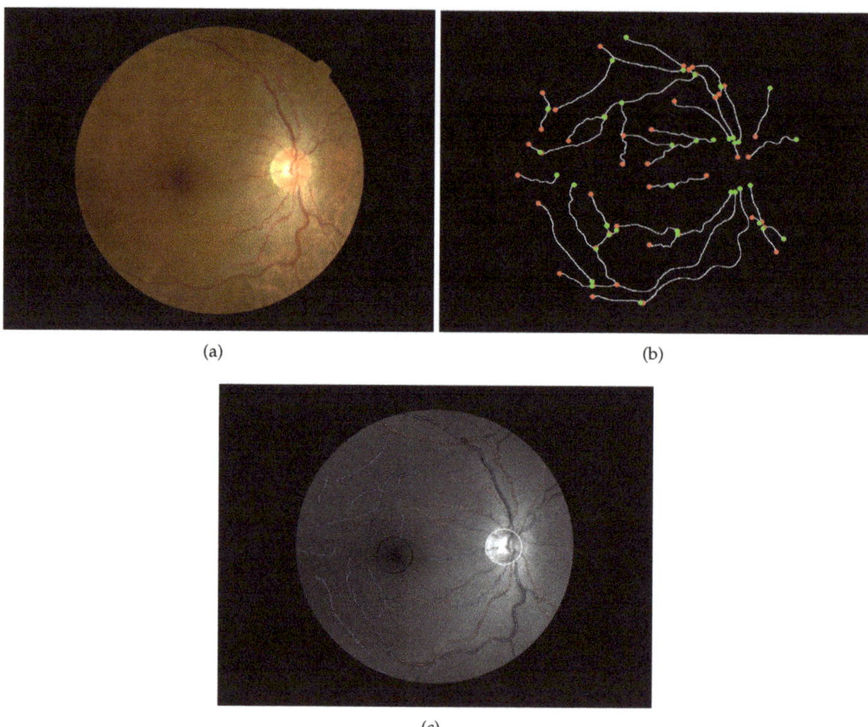

Figure 1. (**a**) Original fundus image. (**b**) Retinal arterio-venous tree extraction and vessel decomposition. (**c**) Domain-related information including the vessel caliber, the distance to the optic disc, the distance to the fovea and the distinction between arteries and veins.

3. Results and Conclusions

In order to evaluate the impact of the integration of domain-related information for the tortuosity characterization, the prognostic performance of the proposed metric was compared to the effectiveness of the baseline metric of reference, based exclusively on a mathematical representation of the vessel segments. The evaluation was performed over a dataset composed by 200 fundus images manually annotated by a group of five experts according to the binary relevant/non-relevant classification. The AUC achieved by the proposed metric was 93.8%, whereas the AUC of the baseline metric exclusively based on mathematical properties was 89.8%. The results show that the incorporation of the anatomical factors allows a reliable characterization of the retinal vascular tortuosity that better represents the specialists' perception.

Author Contributions: L.R., J.N., J.R. and M.O. made substantial contributions to conception and design of the retinal vascular tortuosity assessment, analysis of data and interpretation of the results. S.R. and M.D.A. were involved in the acquisition of the dataset, the rating procedure and the identification of the relevant domain-related information. L.R. performed the experiments and drafted the manuscript and all the authors participated in its critical revision and final approval. All authors have read and agreed to the published version of the manuscript.

Funding: This work is supported by the Instituto de Salud Carlos III, Government of Spain and FEDER funds of the European Union through the DTS18/00136 research projects and by the Ministerio de Ciencia, Innovación y Universidades, Government of Spain through the RTI2018-095894-B-I00 research projects; CITIC, Centro de Investigación de Galicia ref. ED431G 2019/01, receives financial support from Consellería de Educación, Universidade e Formación Profesional, Xunta de Galicia, through the ERDF (80%) and Secretaría Xeral de Universidades (20%).

Conflicts of Interest: The authors declare no conflict of interest. The funders had no role in the design of the study; in the collection, analyses, or interpretation of data; in the writing of the manuscript, or in the decision to publish the results.

References

1. Ramos, L.; Novo, J.; Rouco, J.; Romeo, S.; Álvarez, M.D.; Ortega, M. Computational assessment of the retinal vascular tortuosity integrating domain-related information. *Sci. Rep.* **2019**, *9*, 19940.
2. Grisan, E.; Foracchia, M.; Ruggeri, A. A novel method for the automatic grading of retinal vessel tortuosity. *IEEE Trans. Med. Imaging* **2008**, *27*, 310–319.

Proceedings

Data Extraction in Insurance Photo-Inspections Using Computer Vision †

Mateo Gende [1,2,*], Marcos Ortega [1,2] and Manuel G. Penedo [1,2]

1 Department of Computer Science, University of A Coruña, 15071 A Coruña, Spain; m.ortega@udc.es (M.O.); manuel.gpenedo@udc.es (M.G.P.)
2 CITIC-Research Center of Information and Communication Technologies, University of A Coruña, 15071 A Coruña, Spain
* Correspondence: m.gende@udc.es; Tel.: +34-981-16-70-00 (ext. 5522)
† Presented at the 3rd XoveTIC Conference, A Coruña, Spain, 8–9 October 2020.

Published: 21 August 2020

Abstract: Recent advances in computer vision and artificial intelligence allow for a better processing of complex information in many fields of human activity. One such field is vehicle expertise and inspection. This paper presents the development of systems for the automatic reading of French and Spanish license plates, as well as odometer value reading in dashboard photographs. These were trained and validated with real examples of more than 4000 vehicles, while addressing typical problems with irregular data acquisition. The systems proposed have found use in a real environment and are employed as assistance in vehicle appraisal.

Keywords: computer vision; machine learning; deep learning; license plate; odometer

---

1. Introduction

The ever-advancing field of artificial intelligence is constantly finding new uses in many activities and businesses. Vehicle insurance and expertise companies involve activities such as automobile inspections in which many images are taken to record and verify not only possible damages but also identity, in the form of license plate and vehicle identification number, and usage, like travelled distance. Possible mistakes in acquiring these data can result in problems ranging from insurance fraud allegations to uninsured vehicles being involved in accidents. Experts performing routine activities such as manually typing data in the field are susceptible to these kinds of mistakes. Computer vision techniques, combined with machine learning models, can be used to produce systems that are able to extract this information from images directly. This can have the benefit of speeding up the task of data acquisition by skipping manual typing entirely or, at least, provide a layer of verification by notifying the expert when the data they introduced does not match the automatic reading, indicating there may have been a typing mistake or a problem with how the image was taken. Blurred, poorly illuminated images make for deficient expertise material.

2. Methodology

2.1. License Plate Reading

To read license plate numbers, a hybrid system was developed. First, the license plate is localised and segmented from the image using a novel convolutional neural network based on U-net [1] architecture. From the segmentation map, the four corner points of the license plate rectangle are obtained so a projection transformation can be applied. With this transformation we obtain a flattened, standardised image of just the license plate, from which we can extract the characters using computer vision techniques. Figure 1 shows a summary of these steps. Each character image is then classified using a convolutional neural network. Finally, the ordered character list is filtered using regular expressions to fix possible errors and classified as a Spanish, French or incorrectly read license plate.

Figure 1. Summary of the license plate reading procedure: (**left**) Original image. (**center**) Segmentation and perspective estimation. (**right**) Cleaning, thresholding and single character segmentation.

2.2. Odometer Reading

For odometers, a different approach was taken due to the inherent variability of analysing many different dashboards from any car make or model. To read the kilometre count, the user is asked to mark a point in the image over the digits that they want read. This way the system can limit the surface of the image that is analysed, as well as know which transcription is the one of interest, avoiding unnecessary processing and returning only what was requested from an image that may contain irrelevant data from the trip counter to time and date. A standardised window is taken from the image around the marked point and fed to a YOLOv3 [2] network. This model segments and classifies the numbers found in the image fitting a bounding box for each, as seen on Figure 2. Next, the user-supplied point is used to determine the extent of the figure of interest, including similar numbers in close proximity. For this, the size and position of already included numbers is used, providing flexibility to compensate for much of the heterogeneity of these images.

Figure 2. Example of character recognition in a car odometer.

3. Results

A test dataset of 251 real images of varying illumination and quality was constructed in order to evaluate license plate reading. Of these, 236 were perfectly transcribed by the system, achieving an accuracy of 94.02%. For the remaining 15 incorrectly read, the system was able to report an image quality problem in 12 of them, while three mistakes, 1.20%, went unnoticed. In total, 100 dashboard images were used to test the odometer reading. The system read 75 of them correctly for an accuracy of 75%.

Funding: This research was co-funded by Ministerio de Ciencia, Innovación y Universidades and the European Union (European Regional Development Fund—ERDF) as part of the RTC-2017-5908-7 research project. We wish to acknowledge the support received from the Centro de Investigación de Galicia CITIC, funded by Consellería de Educación, Universidade e Formación Profesional from Xunta de Galicia and European Union. (European Regional Development Fund - FEDER Galicia 2014-2020 Program) by grant ED431G 2019/01.

Acknowledgments: This work was supported by and used material from project STEPS (RTC-2017-5908-7).

Conflicts of Interest: The authors declare no conflict of interest.

References

1. Ronneberger, O.; Fischer, P.; Brox, T. U-net: Convolutional networks for biomedical image segmentation. In *Medical Image Computing and Computer-Assisted Intervention, Proceedings of the MICCAI 2015, 18th International Conference, Part III, Munich, Germany, 5–9 October 2015*; Navab, N., Hornegger, J., Wells, W.M., Frangi, A.F., Eds.; Springer International Publishing: Cham, Switzerland, 2015; pp. 234–241.
2. Redmon, J.; Farhadi, A. YOLOv3: An Incremental Improvement. *arXiv* **2018**, *4*, arXiv:1804.02767.

Proceedings

Probiotic: First Prescriptive Application of Probiotics in Spain †

Sara Alvarez-Gonzalez *, Jose Rodriguez-Fernandez, Ana Belen Porto-Pazos, Alejandro Pazos and Francisco Cedron

Computer Science and Information Technology Department, Research Center on Information and Communication Technologies, University of A Coruña, 15071 A Coruña, Spain; jose.rodriguez3@udc.es (J.R.-F.); ana.portop@udc.es (A.B.P.-P.); alejandro.pazos@udc.es (A.P.); francisco.cedron@udc.es (F.C.)

* Correspondence: sara.alvarezg@udc.es; Tel.: +34-881-01-2666

† Presented at the 3rd XoveTIC Conference, A Coruña, Spain, 8–9 October 2020.

Published: 21 August 2020

Abstract: The study of the intestinal microbiota is one of the greatest challenges in today's clinical environment. Thus, probiotics have been established as a focus for its stability, as they play a key role in its regulation. The development of an automated technique that allows the practitioners the smooth search for the optimal probiotic is postulated as the main objective of this study. Despite the existence of previous attempts at applications for this purpose, they have only been carried out for the countries of origin, preventing them from being used in others such as Spain. Therefore, a system has been developed with open, multi-platform, and free technologies, which manages to locate the optimal probiotic for each pathology.

Keywords: probiotic; microbiota; software

1. Introduction

The human microbiome is the population of all microorganisms with their genes and metabolites that colonize the body. The highest concentration of microbes is found in the intestine and is called intestinal microbiota. The flora of the microbiome has a significant impact on human health, and understanding its effects is one of the most salient challenges facing clinical care nowadays. Intake of the right amount of the appropriate microorganisms can help regenerate the composition and concentration of the microbiota, thus eliminating problems such as dysbiosis [1]. Moreover, bearing in mind the current situation with the Covid-19 pandemic disease, it has been highlighted how alterations in the microbiota may be linked to individual severity with which each patient is affected [2]. A detailed report of the SEPyP (Probióticos, prebióticos y salud) clarified that for a microorganism to be qualified as probiotic it is necessary to scientifically demonstrate that it produces beneficial effects on human health [3]. Furthermore, pharmacists and doctors themselves have started to use probiotics for certain treatments. The data and studies that have been published with clinical results for probiotics strains are beginning to be abundant to such an extent that a professional in the field cannot afford to know all the probiotic products with the clinical evidence that exists.

2. Objective

The main objective of the present work is to simplify the consultation process, so that with a simple, agile, and intuitive application, the search and consultation will be quick and effective, so that the doctor or pharmacist will have all the information available to treat and advise his patients, easily selecting the suitable product, dose, and format for a particular indication.

3. State-of-the-Art

Undeniably, the healthcare world is evolving towards a more technological environment where there are more and more applications that support clinicians in diagnosing pathologies, making decisions, and even prescribing medical treatment to ensure that the medication and its dosage is indeed recommended. The Clinical Guide to Probiotic Product application is a guide to the scientific evidence available for probiotic products that was created with published data from clinical studies for various probiotic strains. Currently, there are two released for products available and supported in Canada and United States [4,5]. Besides the information being extremely worthwhile, due to differences in the commercial products in Spain, it lacks usability.

4. Methodology

This paper presents a prescription assistance system that has been developed with the aim of providing a tool that includes the products endorsed in Spain for later consultation. The system developed has a webpage that enables remote consultation of the appropriate probiotics for a specific pathology in a very simple way (Figure 1a). Once the pathology has been selected, the available probiotics are shown, ordered by the level of evidence. In addition, the system designed is not intended to be static and has an administration panel that allows you to create, edit, or delete information of all kinds. It is also possible to access quick actions to view details, edit, or delete (Figure 1b).

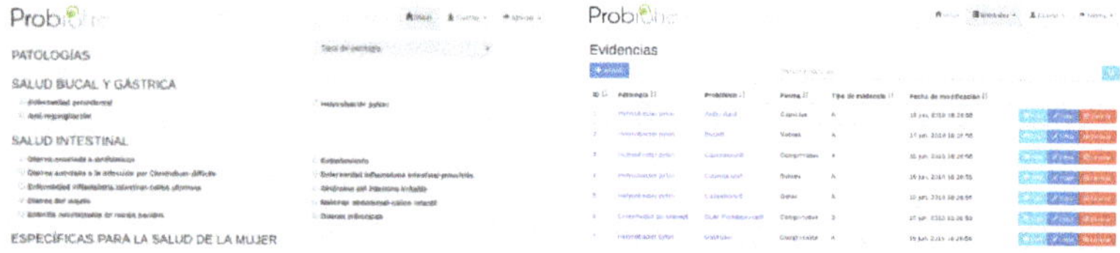

(a) Homepage **(b)** List of evidence of probiotic

Figure 1. Application developed.

Funding: This work has received financial support from the Xunta de Galicia and the European Union (European Social Fund (ESF)).

Acknowledgments: This project was also supported by the General Directorate of Culture, Education and University Management of Xunta de Galicia (Ref. ED431G/01, ED431D 2017/16), the "Galician Network for Colorectal Cancer Research" (Ref. ED431D 2017/23).

References

1. Myers, S. P. The causes of intestinal dysbiosis: A review. *Altern. Med. Rev.* **2004**, *9*, 180–197.
2. Zuo, T.; Zhang, F.; Lui, G.C.; Yeoh, Y.K.; Li, A.Y.; Zhan, H.; Wan, Y.; Chung, A.; Cheung, C.P.; Lai, C.K.; et al. Alterations in Gut Microbiota of patients with COVID-19 during time of hospitalization. *Gastroenterology* **2020**. doi:10.1053/j.gastro.2020.05.048
3. SEPyP, Probióticos, Prebióticos y Salud: Evidencia Científica. Available online: http://ergon.es/wp-content/uploads/2016/07/ergon_primeras_Manual_SEPyP.pdf (accessed on 20 July 2020).
4. Skokovic-Sunjic, D. Clinical Guide to Probiotic Products Available in Canada: 2020 Edition. 2020. Available online: http://probioticchart.ca(accessed on 20 July 2020).
5. Skokovic-Sunjic, D. Clinical Guide to Probiotic Products Available in the US: 2020 Edition. 2020. Available online: http://probioticchart.ca (accessed on 20 July 2020).

Proceedings

Deep Image Segmentation for Breast Keypoint Detection †

Tiago Gonçalves [1,*], Wilson Silva [1], Maria J. Cardoso [2] and Jaime S. Cardoso [1]

1 INESC TEC and Faculdade de Engenharia, Universidade do Porto, 4200-465 Porto, Portugal; wilson.j.silva@inesctec.pt (W.S.); jaime.cardoso@inesctec.pt (J.S.C.)
2 Champalimaud Foundation and Nova Medical School, 1400-038 Lisboa, Portugal; maria.joao.cardoso@fundacaochampalimaud.pt
* Correspondence: tiago.f.goncalves@inesctec.pt

† Presented at the 3rd XoveTIC Conference, A Coruña, Spain, 8–9 October 2020.

Published: 21 August 2020

Abstract: The main aim of breast cancer conservative treatment is the optimisation of the aesthetic outcome and, implicitly, women's quality of life, without jeopardising local cancer control and overall survival. Moreover, there has been an effort to try to define an optimal tool capable of performing the aesthetic evaluation of breast cancer conservative treatment outcomes. Recently, a deep learning algorithm that integrates the learning of keypoints' probability maps in the loss function as a regularisation term for the robust learning of the keypoint localisation has been proposed. However, it achieves the best results when used in cooperation with a shortest-path algorithm that models images as graphs. In this work, we analysed a novel algorithm based on the interaction of deep image segmentation and deep keypoint detection models capable of improving both state-of-the-art performance and execution-time on the breast keypoint detection task.

Keywords: aesthetic assessment of breast cancer surgery outcomes; artificial intelligence; breast cancer; breast cancer conservative treatment; computer vision; deep learning; image segmentation; machine learning; keypoint detection

---

1. Introduction

Breast cancer is a highly mutable and rapidly evolving disease, however, thanks to the generalised use of breast cancer screening and better treatments, approximately 90% of the cases can be cured [1,2]. Therefore, it is now possible to employ breast cancer conservative treatment (BCCT) approaches, which only require the removal of cancerous tissue with a rim of healthy tissue, instead of radical mastectomy-based approaches, which require the removal of the entire breast and posterior breast reconstruction [3]. In both cases, it is possible to obtain good cosmetic results, and consequently, improve patients' quality of life. The objective assessment of the cosmetic results, which acts as a reliable proxy of the quality of the treatment and valuable input to the improvement of current techniques, is performed through the analysis of digital photographs, from which, several features are extracted and used to train a classification algorithm [4]. To facilitate the extraction of such features, the annotation of several breast keypoints is required. Several semi-automatic methods to perform this keypoint annotation are already available, however, they still need the input from the user and are time-consuming and computationally demanding. Recently, Silva et al. showed that deep learning algorithms may be part of the answer. They introduced a deep algorithm based deep neural networks (DNN) which receives an image as input and returns the coordinates of the breast keypoints as output [5], which are then given to a shortest-path algorithm that models images as graphs to refine breast keypoint localisation. Although this increased the performance of the task of breast keypoint annotation, this is still computationally complex. To overcome this issue, we proposed a novel deep keypoint detection algorithm, which combines the approach by Silva et al. together with deep image segmentation model that refines breast keypoint localisation in lesser time, and with improved precision [6].

2. Materials and Methods

Our approach combines the DNN proposed by Silva et al. and a deep image segmentation model, U-Net++ [7], in a pipeline (see Figure 1). The intuition behind this approach is that it is easier to detect breast contours if one is capable to detect breasts first. We started by the training of both the U-Net++ and the DNN by Silva et al. in the breast detection (i.e., generate breast masks) and in the keypoint detection tasks, respectively. The complete pipeline works as follows: the image is given as input to U-Net++, which returns a breast segmentation mask; from this mask, contours are extracted using the marching squares algorithm [8]; the image is given as input to the DNN by Silva et al., which returns a set of breast keypoints; the refined localisation of breast keypoints is obtained with the projection of this set onto the breast segmentation mask contours through the minimisation of the Euclidean Distance of a given breast keypoint and a given contour (processing step). We also trained both the DNN and the complete keypoint method proposed by Silva et al. for comparison purposes. All the experiments were performed taking into account 5-fold cross-validation. In addition to the study of algorithms' performance, a study on the algorithms' execution time, in seconds, was also done, regarding the interest in the deployment of these algorithms into a web-application for both the research and medical communities.

Figure 1. Proposed deep image segmentation keypoint detection method [6].

3. Results

Table 1 presents the average error distance (measured in pixels) and the average execution time (measured in seconds) of each model inference on the test set. Figure 2 shows a visual example of the obtained results.

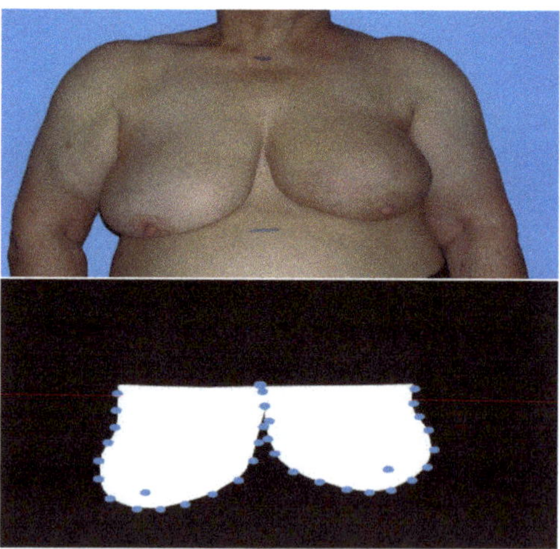

Figure 2. Example of the results obtained with the proposed deep image segmentation method. The first image is the input photograph and the second image is the U-Net++ predicted mask with the detected breast keypoints (after the processing step).

Table 1. Average error distance for endpoints, breast contours and nipples, measured in pixels and average execution time of the models' inferences. Best results are highlighted in bold. Note: STD stands for standard deviation and Max stands for maximum error.

Model	Endpoints			Breast Contours			Nipples			Execution Time (s)
	Mean	STD	Max	Mean	STD	Max	Mean	STD	Max	
Silva et al. Keypoint Detection DNN	40	**33**	**182**	21	8	72	**70**	**39**	**218**	**150**
Silva et al. Keypoint Detection Method	40	**33**	**182**	13	14	104	**70**	**39**	**218**	1704
Proposed Keypoint Detection Method	**38**	34	195	**11**	**5**	**34**	**70**	**39**	**218**	280

4. Discussion

Our keypoint detection method surpassed both the DNN-based keypoint detection and the keypoint detection method from Silva et al. in the endpoints and breast contours detection tasks, which were, to our knowledge, the state-of-the-art breast keypoint detection algorithms. Besides, this novel algorithm achieves lower values of standard deviation and maximum error, which suggests more consistency when compared with the other two. Regarding the study of performance, our keypoint detection method presents the best balance between time-efficiency and accuracy, being the most accurate model, with a time efficiency comparable to the most time-efficient method.

5. Conclusions

In this work, we proposed keypoint detection method that combines a deep keypoint detection and deep image segmentation models, and capable of achieving a good balance between both performance and execution-time. Further studies should be focused on the development of a fully end-to-end deep keypoint detection model, trained with a multi-term loss function, and on the deployment of these breast keypoint detection algorithms into a fully-functional web-application for both the research and medical communities.

Author Contributions: Conceptualization, T.G., W.S. and J.S.C.; methodology, T.G., W.S. and J.S.C.; software, T.G. and W.S.; validation, J.S.C.; formal analysis, J.S.C.; investigation, T.G., W.S., M.J.C. and J.S.C.; resources, M.J.C. and J.S.C.; data curation, M.J.C. and J.S.C.; writing—original draft preparation, T.G.; writing—review and editing, W.S. and J.S.C.; visualization, M.J.C. and J.S.C.; supervision, J.S.C.; project administration, J.S.C.; funding acquisition, J.S.C. All authors have read and agreed to the published version of the manuscript.

Funding: This work was partially financed by the ERDF - European Regional Development Fund through the Operational Programme for Competitiveness and Internationalisation - COMPETE 2020 Programme and by National Funds through the Portuguese funding agency, FCT - Fundação para a Ciência e a Tecnologia within project "POCI-01-0145-FEDER-028857" and PhD grant number SFRH/BD/139468/2018.

Acknowledgments: The authors thank Florian Fitzal for sharing the VIENNA dataset.

Conflicts of Interest: The authors declare no conflict of interest.

References

1. Oliveira, H.P.; Cardoso, J.S.; Magalhaes, A.; Cardoso, M.J. Methods for the Aesthetic Evaluation of Breast Cancer Conservation Treatment: A Technological Review. *Curr. Med Imaging Rev.* **2013**, *9*, 32–46. doi:10.2174/1573405611309010006.
2. Wilkes, G.M.; Barton-Burke, M. *2018 Oncology Nursing Drug Handbook*; OCLC: 1030285714; Jones & Bartlett Learning: Burlington, MA, USA, 2018.
3. Street, W. *Cancer Facts & Figures*; American Cancer Society: Atlanta, GA, USA, 2018; p. 76.
4. Cardoso, J.S.; Cardoso, M.J. Towards an intelligent medical system for the aesthetic evaluation of breast cancer conservative treatment. *Artif. Intell. Med.* **2007**, *40*, 115–126. doi:10.1016/j.artmed.2007.02.007.
5. Silva, W.; Castro, E.; Cardoso, M.J.; Fitzal, F.; Cardoso, J.S. Deep Keypoint Detection for the Aesthetic Evaluation of Breast Cancer Surgery Outcomes. In Proceedings of the IEEE International Symposium on Biomedical Imaging (ISBI'19), Venice, Italy, 8–11 April 2019.

6. Gonçalves, T.; Silva, W.; Cardoso, M.J.; Cardoso, J.S. A novel approach to keypoint detection for the aesthetic evaluation of breast cancer surgery outcomes. In *Health and Technology*; Springer: Berlin/Heidelberg, Germany, 2020; pp. 1–13.
7. Zhou, Z.; Rahman Siddiquee, M.M.; Tajbakhsh, N.; Liang, J. UNet++: A Nested U-Net Architecture for Medical Image Segmentation. In *Deep Learning in Medical Image Analysis and Multimodal Learning for Clinical Decision Support*; Stoyanov, D., Taylor, Z., Carneiro, G., Syeda-Mahmood, T., Martel, A., Maier-Hein, L., Tavares, J.M.R., Bradley, A., Papa, J.P., Belagiannis, V., et al., Eds.; Springer International Publishing: Cham, Switzerland, 2018; Volume 11045, pp. 3–11. doi:10.1007/978-3-030-00889-5_1.
8. Lorensen, W.E.; Cline, H.E. Marching cubes: A high resolution 3D surface construction algorithm. In Proceedings of the 14th Annual Conference on Computer GRAPHICS and interactive Techniques—SIGGRAPH '87; ACM Press: New York, NY, USA, 1987; pp. 163–169. doi:10.1145/37401.37422.

Proceedings

An Intelligent and Collaborative Multiagent System in a 3D Environment †

Alejandro Rodríguez-Arias *, Bertha Guijarro-Berdiñas and Noelia Sánchez-Maroño

Campus de Elviña s/n, Universidade da Coruña, CITIC, 15071 A Coruña, Spain; berta.guijarro@udc.es (B.G.-B.); noelia.sanchez@udc.es (N.S.-M.)

* Correspondence: alejandro.rodriguez.arias@udc.es

† Presented at the 3rd XoveTIC Conference, A Coruña, Spain, 8–9 October 2020.

Published: 21 August 2020

Abstract: Multiagent systems (MASs) allow facing complex, heterogeneous, distributed problems difficult to solve by only one software agent. The world of video games provides problems and suitable environments for the use of MAS. In the field of games, Unity is one of the most used engines and allows the development of intelligent agents in virtual environments. However, although Unity allows working in multiagent environments, it does not provide functionalities to facilitate the development of MAS. The aim of this work is to create a multiagent system in Unity. For this purpose, a predator–prey problem was designed in which the agents must cooperate to arrest a thief driven by a human player. To solve this cooperative problem, it is required to create the representation of the environment and the agents in 3D; to equip the agents with vision, contact, and sound sensors to perceive the environment; to implement the agents' behaviors; and, finally but not less important, to build a communication system between agents that allows negotiation, collaboration, and cooperation between them to create a complex, role-based chasing strategy.

Keywords: intelligent agents; multiagent systems; unity; agent communication language

1. Introduction

Multiagent systems (MASs) are an artificial intelligence discipline that allows facing complex, heterogeneous, and distributed problems that cannot be solved by a single software agent. A MAS is a society of homogeneous or heterogeneous intelligent agents specialized in solving part of a problem who cooperate or compete for resources to achieve their goals [1]. In a MAS, each agent acts autonomously. Therefore, agents need to be able to negotiate, cooperate, and coordinate to achieve their goals [2]. This is only possible if some agent communication language (ACL) has been developed.

The world of video games provides problems and suitable environments for the use of MAS. Unity is one of the most used video game engines and allows the development of intelligent agents in virtual 2D/3D environments [3]. The Unity Machine Learning (ML) Agents project enables games and simulations run in Unity to be used as intelligent agent training environments using machine learning algorithms, enabling developers and researchers to work with Unity and machine learning [4]. Even though both Unity and ML Agents allow working in environments with multiple agents, none provide functionalities to work with MAS, that is, there are no tools for agents to communicate and, consequently, they cannot carry out tasks in a coordinated way.

The objective of this work is to create a multiagent system in Unity. To test our development, we use a predator–prey problem where the MAS, in the role of security guards, must use a collaborative strategy that allows it to catch a human player, in the role of a thief, who has stolen a treasure and escapes using the exit door in a virtual 3D environment.

2. New Functionalities for the Development of Multiagent Systems in Unity

For solving problems, an agent must be able to perceive the environment and act in consequence. In the case of complex problems in which the actions of several agents are necessary, communication actions are also mandatory to permit coordination and/or negotiation between agents. Therefore, it is necessary to build a communication system in Unity that allows this communication.

2.1. Single Agents

In our system, each agent has been endowed with three sensors that allows them to perceive the environment they are in (in our example, mainly walls, doors, and the player). These three sensors are: (1) a vision sensor that allows the agents to recognize any object in their front view cone; (2) a sound sensor that allows the agents to hear, through a nearby radio, the noise of the steps generated by the player; (3) a contact sensor that allows the agents to identify the objects that they collide with.

These agents can perform three different actions: (1) move freely around the stage using the A* algorithm to find the optimal route to their objective; (2) open/close doors to go through them; (3) capture the player once an agent collides with him.

2.2. The Agent Communication System

A communication system between agents is based on three key points: (1) the structure of messages; (2) a system for sending, creating, and receiving messages; and (3) a system for managing conversations between agents.

For the structure of the messages, it is necessary to use an agent communication language (ACL). An ACL is a standard that provides high-level methods for an agent to exchange information. The ACL itself defines the types and meanings of the messages that agents can exchange. Most ACLs are based on speech act theory [5], which means that messages are communicative acts.

As Unity lacks a communication language between agents, we developed a plugin based on the ACL proposed by the Foundation for Intelligent Physical Agents (FIPA) [6,7], reducing the number of communicative acts and parameters available in FIPA-ACL to those essential for communication and developing a collaborative strategy complex enough to solve problems.

2.3. Behavior

In our system, the agents' behavior varies over time. Initially, everyone is assigned a limited area where they patrol. When one detects the player, the agents form a team and each one assumes a role according to their current position with respect to the player, other agents, and the player's objective.

An agent can take on different roles: a *goalkeeper*, who patrols near the player's target; *defense*, attempting to block accessible rooms from the player's current position; an *attacker*, who will attempt to block exits from the player's current room; and a *pursuer*, who will chase the player across the scenario.

Agents with the same role form a team with randomly assigned leader who helps coordinate negotiations on the objective of each agent in their role (for example, which door access will be blocked). Thanks to the designed communication system, each agent maintains negotiations with other agents with the same role and shares information with all of the agents in the system to try to use the most effective strategy at all times.

An example of how the system works can be seen in https://youtu.be/CEw7XkMlKfc

3. Conclusions

In this work we have provided Unity and Unity ML Agents with the necessary facilities to develop a MAS, that is: (a) a communication system; (b) an ACL based on FIPA to understand the interchanged messages; (c) basic sensors to allow agents to perceive the environment. These functionalities allow the resolution of problems that require collaborative solutions between agents, as demonstrated in the predator–prey problem implemented.

Author Contributions: These authors contributed equally to this work.

Funding: This work has been supported in part by the Xunta de Galicia (Grants ED431C 2018/34 and ED431G/01), the European Union's H2020 SMARTEES project (Grant 763912), and the European Union FEDER funds.

References

1. Sycara, K. Multiagent Systems. *AI Magazine*, 15 June 1998.
2. Wooldridge, M. *An Introduction to Multiagent Systems*; John Wiley & Sons: Hoboken, NJ, USA, 2009.
3. Toftedahl, M.; Engström, H. A Taxonomy of Game Engines and the Tools that Drive the Industry. In Proceedings of the Digra International Conference 2019, Game, Play and the Emerging Ludo-Mix, Kyoto, Japan, 6–10 August 2019.
4. Juliani, A.; Berges, V.P.; Vckay, E.; Gao, Y.; Henry, H.; Mattar, M.; Lange, D. Unity: A general platform for intelligent agents. *arXiv* **2018**, arXiv:1809.02627.
5. Austin, J.L. *How to Do Things with Words*; Oxford University Press: Oxford, UK, 1975; Volume 88.
6. Fipa, A. *Fipa Acl Message Structure Specification*; Foundation for Intelligent Physical Agents: Geneva, Switzerland, 2002.
7. FIPA, T. Fipa communicative act library specification. Avaliable online: https://www.fipa.org/specs/fipa00037/PC00037E.html (accessed on 30 November 2000).

Proceedings

Comparison of Image Compressions: Analog Transformations †

Jose Balsa

CITIC Research Center, Universidade da Coruña (University of A Coruña), 15071 A Coruña, Spain; j.balsa@udc.es

† Presented at the 3rd XoveTIC Conference, A Coruña, Spain, 8–9 October 2020.

Published: 21 August 2020

Abstract: A comparison between the four most used transforms, the discrete Fourier transform (DFT), discrete cosine transform (DCT), the Walsh–Hadamard transform (WHT) and the Haar-wavelet transform (DWT), for the transmission of analog images, varying their compression and comparing their quality, is presented. Additionally, performance tests are done for different levels of white Gaussian additive noise.

Keywords: analog image transformation; analog image compression; analog image quality

---

1. Introduction

Digitized image coding systems employ reversible mathematical transformations. These transformations change values and function domains in order to rearrange information in a way that condenses information important to human vision [1]. In the new domain, it is possible to filter out relevant information and discard information that is irrelevant or of lesser importance for image quality [2]. Both digital and analog systems use the same transformations in source coding. Some examples of digital systems that employ these transformations are JPEG, M-JPEG, JPEG2000, MPEG-1, 2, 3 and 4, DV and HDV, among others. Although digital systems after transformation and filtering make use of digital lossless compression techniques, such as Huffman.

In this work, we aim to make a comparison of the most commonly used transformations in state-of-the-art image compression systems. Typically, the transformations used to compress analog images work either on the entire image or on regions of the image. The transforms whose performance will be analyzed with compression and analog noise are the discrete Fourier transform (DFT), discrete cosine transform (DCT), the Walsh–Hadamard transform (WHT) and the Haar-wavelet transform (DWT) [3,4].

These transformations can be applied to the whole image or to parts of the image. The decision will depend on the performance of the transformation, the computational load, the desired compression level and the impact of the noise on the coefficients of the transformed domain. In this work, we intend to make an analysis of the performance of the above-mentioned transformations by varying the characteristics listed.

2. System Description

The system proposed and implemented in MATLAB performs the four transformations, DFT, DCT, WHT and the Haar-based wavelet (DWT), simultaneously, with the same images, the same compression ratio and additive white Gaussian noise (AWGN) with the same intensity. The purpose of the system is to test and compare the performance of the four transformations applied to images as a function of the compression ratio and noise.

The operation of the DFT, DCT and WHT transforms are similar. These transforms change the domain of the image symbols to concentrate the relevant information into certain coefficients, allowing for compression by setting a transmission threshold. These transforms are applied to the entire image or to smaller divisions of the image, usually in squares. The selection of the size of the block where it will be applied is also relevant to the quality of the image [1], and we will take it into account in the results.

The Haar-based wavelet [4] is a transformation that condenses the important information of an image onto certain symbols. The reverse transformation allows the recovery of the exact image. For the compression, a threshold is set from which the symbols are approximated by zero and it is these zeros over the total symbols that allow the calculation of the compression ratio.

Therefore, the system created allows a comparison of the four transformations as a function of image quality, using the structural similarity index measure (SSIM) [5], and varying the compression ratio and noise intensity. A bank of 14 images with different characteristics is used.

3. Results and Conclusions

The results shown are for a block size of 16 × 16 for the DFT, DCT and WHT transformations. In the case of the Haar-based wavelet transformation, it is performed on the whole image. In addition, a filling with zeros is performed if the multiplicity is not met, as in the case of WHT and the DWT. For this purpose, it is necessary to do the reverse transform at the reception and then remove the image positions that were filled with zeros.

Figure 1a shows the result for when there is practically no noise; in this case, we can see how the DWT has the best behavior, followed by the DCT. It should be noted that the DFT is sluggish and does not achieve the maximum quality of 1 of SSIM. This is due to the fact that when we do the DFT transform we get twice as many symbols as the rest of the transforms because they are in the real and imaginary domain, so this coding is considerably penalized when comparing the number of symbols transmitted.

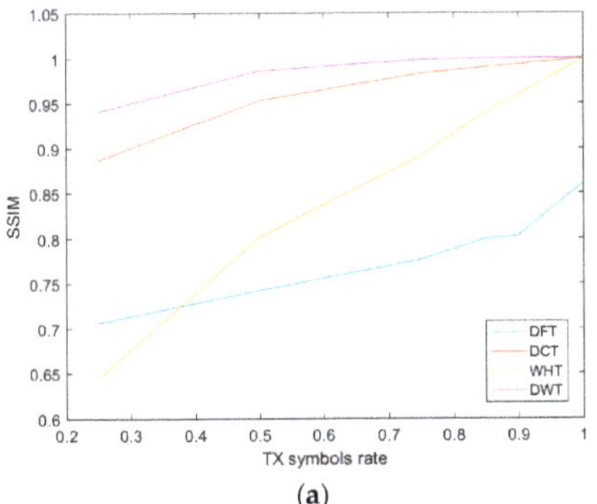

(a)

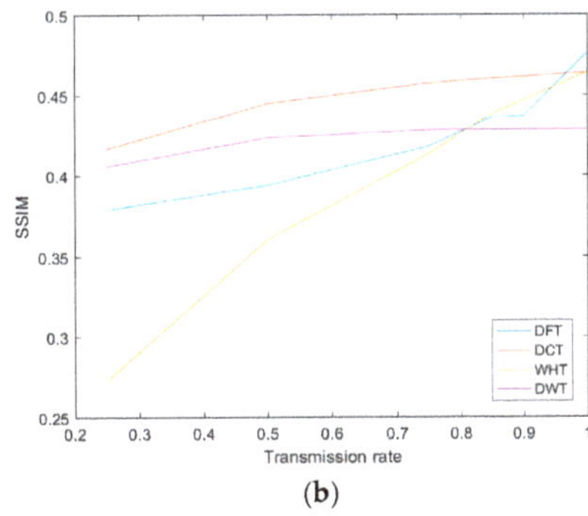

(b)

Figure 1. The relationship between compression rate and image quality (SSIM) is shown for two levels of signal-noise rate (SNR): (**a**) 100 dB additive white Gaussian noise (AWGN); (**b**) 15 dB AWGN.

Figure 1b shows the result for the above situation, but with an AWGN noise level of 15 dB. This time, a better performance of the DCT is shown in general. It can be noticed that the DFT, when there is no compression, obtains better results than the rest of the transformations, but we must emphasize that we are in a very small range of SSIM values, so its relevance is small.

As a conclusion, we can say that, among the most known transformations for images, those that present a greater performance are the DCT and the wavelet. In view of the results obtained, the DCT presents a higher quality of SSIM for noise levels below 20 dB AWGN and the wavelet for higher levels of 20 dB. With this, we can say that it is interesting to use the wavelets for digital systems because the noise does not affect the transformed values. However, for analog systems, where noise is introduced into the transformed symbols, the DCT is the best choice because, in view of the results, it offers a better balance between noise-free and noisy transmissions. Finally, we can say that the WHT and the DFT offer the worst results for transmissions with and without noise.

References

1. Pratt, W.K. *Digital Image Processing*; John Wiley & Sons, Ltd.: New York, NY, USA, 2006.
2. Netravali, A.N.; Limb, J.O. Picture coding: A review. *Proc. IEEE* **1980**, *68*, 366–406.
3. Gonzalez, R.C.; Woods, R.E. *Digital Image Processing*; Pearson, Prentice Hall: Upper Saddle River, NJ, USA, 2008.
4. Talukder, K.H.; Harada, K. Haar wavelet based approach for image compression and quality assessment of compressed image. *arXiv* **2010**, arXiv:1010.4084.
5. Wang, Z.; Bovik, A.C.; Sheikh, H.R.; Simoncelli, E.P. Image quality assessment: From error visibility to structural similarity. *IEEE Trans. Image Process.* **2004**, *13*, 600–612, doi:10.1109/TIP.2003.819861.

Proceedings

Adaptive Real-Time Method for Anomaly Detection Using Machine Learning †

David Novoa-Paradela *, Óscar Fontenla-Romero and Bertha Guijarro-Berdiñas

CITIC Research Center, Universidade da Coruña, 15071 A Coruña, Spain; oscar.fontenla@udc.es (Ó.F.-R.); berta.guijarro@udc.es (B.G.-B.)

* Correspondence: david.novoa@udc.es

† Presented at the 3rd XoveTIC Conference, A Coruña, Spain, 8–9 October 2020.

Published: 22 August 2020

Abstract: Anomaly detection is a sub-area of machine learning that deals with the development of methods to distinguish among normal and anomalous data. Due to the frequent use of anomaly-detection systems in monitoring and the lack of methods capable of learning in real time, this research presents a new method that provides such online adaptability. The method bases its operation on the properties of scaled convex hulls. It begins building a convex hull, using a minimum set of data, that is adapted and subdivided along time to accurately fit the boundary of the normal class data. The model has online learning ability and its execution can be carried out in a distributed and parallel way, all of them interesting advantages when dealing with big datasets. The method has been compared to other state-of-the-art algorithms demonstrating its effectiveness.

Keywords: anomaly detection; convex hull; data streaming; big data

---

1. Introduction

Anomaly detection, also known as one-class classification, is the process of identifying unexpected events in data that differ from what is considered normal instances. It has two basic assumptions: anomalies only occur very rarely and their features differ from the normal data significantly. Because anomalous data occurs very sporadically, in most real-world problems only data of the normal class is available. Then, most machine learning approaches for two-class supervised classification are not applicable to develop automatic methods for anomaly detection. Therefore, it requires specific machine learning methods whose training phase is generally carried out using only normal data. These methods try to model the normal class boundaries, so that new data can be classified by checking whether they belong to the normal class or not. This type of problem is frequent in real-world scenarios, such as predictive maintenance of industrial machinery. In this of problems, the ability to learn in real time can be essential as there may not be a sufficient amount of data at the beginning of the learning process but over time. There are many use cases (medical, IT security, etc.) where this situation happens and the detection methods must be able to start making decisions as soon as possible with very little initial knowledge and adapt this knowledge as new data are available. This paper presents the adaptation of an anomaly-detection method [1,2] based on convex hulls and random projections. The main contributions are to allow the convex hulls to be dynamically changed in an online learning scenario and to represent non-convex regions as a union of several convex hulls. The limits of the normal class will adapt when new data is processed, without store all the data or retrain from scratch.

2. Base Methods

Calculating the CH (Convex Hull) in high-dimensional spaces is a computationally expensive task. Due to this, several anomaly-detection methods choose to project the data on 2-D spaces in which the calculation of CHs is simple [1]. This random projection technique is based on the idea that high-dimensional data spaces can be projected into a lower-dimensional space without significantly losing the data structure if multiple projections are used. Base methods consist of the following:

1. Learning phase: Given a data set (normal data), a number τ of random projections of the data e are made onto 2-D subspaces. First, τ random matrices are generated. Second, the training set is projected into the space generated by each projection matrix. Finally, the CH's vertices are calculated in each projection, this being the aim of training (see projections P_1, P_2, P_3 in Figure 1). These vertices are projections of the original data; therefore, the algorithms only need to store the vertices forming the CHs and the projection matrices, discarding the rest of the training data.
2. Classification phase: To predict the class of a new data point, it is first projected using the τ projections generated during the training phase. For each projection, and given the set of vertices of the CH of that 2-D space, it is possible to check if the point is inside the corresponding polygon. It will be classified as normal only if it is inside all CHs. This procedure is shown in Figure 1. The new point (green) will be classified as an anomaly since it falls out of CH in the P_1 projection.

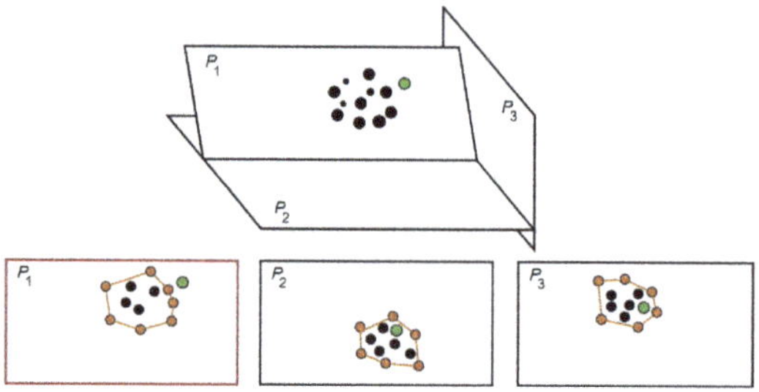

Figure 1. Projections of a 3-D data cloud (black), CHs of each 2-D space (orange) and a new point.

3. Online and Sub-Divisible Distributed Scaled Convex Hull

The main objective of OSHULL (**O**nline and **S**ub-divisible Distributed Scaled Convex **Hull**), is to carry out online one-class learning, so that starting from a minimum set of initial data points, it allows learning and adjusting the model with the arrival of new normal data. Also, because base methods are not suitable for dealing with datasets with non-convex shapes, OSHULL implements this capability. The procedures described below are applied independently to the CH at each of the projections.

3.1. Convex Hull Adjustment

To provide the ability to learn in real time, the limits of CHs must be readjusted as new data is available to the system. To do this, and at the same time guarantee some system stability, these limits will be expanded whenever a considerable number of data falls systematically and concentrated around the same area between the limits of the CH and a margin. In this case, the behavior of the algorithm will be to extend the limits of the CH to cover, as far as possible, that area where normal data did not appear when building the initial CH.

3.2. Region Subdivision

Because representing datasets using a single CH produces poor results in datasets with non-convex geometric shapes, the OSHULL method implements an iterative process that subdivides CHs for a better fit to the shape of the data. In this way, starting from an initial CH, it can be recursively subdivided into as many convex hulls as necessary to properly approximate the shape of normal data. Therefore, the method may represent non-convex regions as the union of various convex regions. At each projection several convex hulls can coexist that will continue to be readjusted individually.

3.3. Freezing Process

In the same way that a CH must be subdivided, a CH with all its edges at low distances from the data, and therefore well adjusted, must be maintained until the end of training and prevent it from being unnecessarily subdivided. This process is called freezing. A CH will freeze if all its edges are at normal distances and it has not been subdivided for a given number of iterations.

3.4. Pruning Process

After several subdivisions, some convex hulls can cover empty regions of normal data. Also, small convex hulls can be found that overlap the margins of other adjacent convex hulls. To get rid of them, a periodic pruning process is performed that eliminates those convex hulls that have not received data inside during a period, as it is assumed that they do not represent normal data.

4. Results

To evaluate the method, 8 datasets were employed (5 artificial and 3 real ones). Artificial datasets were used to evaluate the ability of the method to adapt to certain 3-D shapes. Real datasets were used to evaluate it in higher-dimensional datasets (6, 19 and 30 features). The results obtained by our method in an anomaly-detection scenario were compared to state-of-the-art algorithms working in batch mode. As they do not have the ability to adapt in real time, they were trained with the full training dataset. Conversely, the training process of OSHULL was carried out in real time: first creating the convex hulls with a small dataset and then iteratively readjusting them with the remaining data. Table 1 contains the average results for the test sets. Similarity was used as metric because is a balance between accuracy and recall. Although our method had the fourth position, its average similarity is close to the top, despite the disadvantage of being trained in online mode compared to batch mode.

Table 1. Average similarity (%) ± standard deviations for the different algorithms.

Algorithm	LOF	O-SVM	RC	OSHULL	IF	O-DSCH
Avg. similarity	**91.6± 5.9**	90.6± 6.6	89.4± 8.8	89.3± 7.8	85.1± 6.3	71.7± 19.1

5. Conclusions

OSHULL is an anomaly-detection method that offers online learning without significant loss of performance compared to classic batch methods, and its subdivision capacity favors it for treating non-convex problems. It is a light model as it is not necessary to store all the data for the creation or adaptation of CHs and it just needs to store their vertices. As an additional feature, the convex closures of each projection could be executed in parallel and distributed to achieve greater efficiency.

Author Contributions: These authors contributed equally to this work.

Funding: This work has been supported by Spanish Government's Secretaría de Estado de Investigación (Grant TIN2015-65069-C2-1-R), Xunta de Galicia (Grants ED431C 2018/34 and ED431G/01) and EU FEDER funds.

References

1. Casale, P.; Pujol, O.; Radeva, P. Approximate polytope ensemble for one-class classification. *Pattern Recognit.* **2014**, *47*, 854–864.
2. Fernández-Francos, D.; Fontenla-Romero, O.; Alonso-Betanzos, A. One-class convex hull-based algorithm for classification in distributed environments. *IEEE Trans. Syst. Man Cybern. Syst.* **2018**, *50*, 386–396.

Proceedings

Numerical Model of a 28-GHz Frequency Diverse Array Antenna †

Marc Bernice Angoue Avele *, Julio Brégains, Roberto Maneiro, José A. García-Naya and Luis Castedo

CITIC Research Center, Department of Computer Engineering, Universidade da Coruña, 15071 A Coruña, Spain; julio.bregains@udc.es (J.B.); roberto.maneiro@udc.es (R.M.); jagarcia@udc.es (J.A.G.-N.) luis.castedo@udc.es (L.C.)

* Correspondence: avele.marc@udc.es; Tel: +34-644435147

† Presented at the 3rd XoveTIC Conference, A Coruña, Spain, 8–9 October 2020.

Published: 22 August 2020

Abstract: In this work we make use of the frequency diverse array (FDA) concept, whose design is based upon a frequency increment across the antenna elements to generate a beam steering that is a function of angle, time and range. For a possible use of this technique in 5G detection systems, a 28-GHz FDA numerical model, designed with help of a software tool, is analyzed. Some practical conclusions are drawn from the presented results.

Keywords: frequency diverse array; 5G; numerical simulation

1. Introduction

The next-generation of cellular network (5G) is approaching evolution beyond the mobile Internet to the IoT (Internet of Things) in the near future, with the expectation of communications being available everywhere. 5G is transforming the wireless telecommunication arena, paving the way for the next generation of cellular networks that will include more devices and will enable faster communications through higher speeds. Presumably, the devices technologies will have to adapt to this new approach through the adopted frequency bands. Among many frequency spectra, the millimeter band has been considered as a good candidate for 5G cellular communications, since it provides higher data rate services because of increased channel bandwidth compared to lower frequency bands [1]. A promising antenna technology for utilizing millimeter band communications was introduced [2,3], namely, frequency diverse array (FDA).

FDA is a technology that employs frequency increment across the array elements. With such a technology, beam steering is achieved by applying a linear phase progression across the aperture, creating both a deformation of the antenna pattern in the steering process and a limited scanning.

2. System Description of the Numerical 28-GHz FDA Model

Let us consider an N-element FDA [1,2] whose main axis is aligned along the z axis and whose inter-element distance between contiguous elements is a constant value d (see Figure 1). For a case study, we will set the array by using circular patches whose normals are pointing towards the y axis and that, for the sake of simplicity, will be considered to radiate a $\cos^2\psi$-type power pattern [4]. To further simplify the analysis, the static excitations of the elements will be uniform (i.e., $I_n = 1$, for $n = 0, 1, \ldots, N-1$). Besides, the yz plane ($\varphi = 90°$, see Figure 1) will be taken for representing the main radiation pattern with ψ varying from −90° to 90°. Under those assumptions, the beam power pattern at point $P(r, \psi, 90°)$ will be proportional to [1,2]:

$$P(r,\psi,d,\lambda_c,\Delta f) = \left|\cos(\psi)\sum_{n=0}^{N-1} e^{2j\pi n\left[r\frac{\Delta f}{c}+\frac{d}{\lambda_c}\sin\psi\right]}\right|^2 \quad (1)$$

where c is the speed of light in vacuum, $\lambda_c = c/f_c$ is the wavelength at the carrier (working) frequency f_c, r is the range (distance) measured from the center of the array coordinate system (see Figure 1) and Δf is the frequency shift that produces the range-dependent behavior of the FDA pattern [1,2].

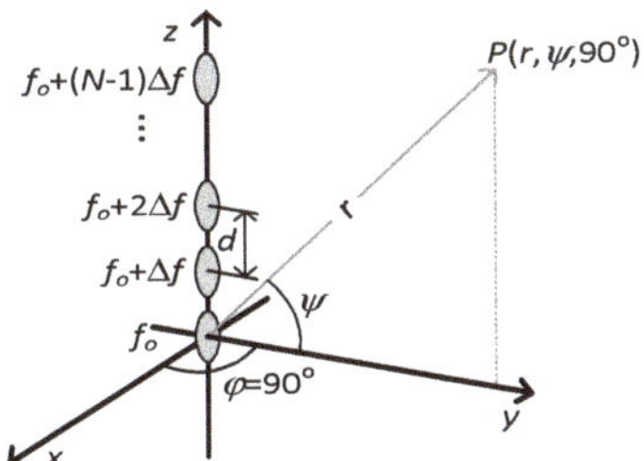

Figure 1. Linear FDA composed of circular patches.

3. Results and Future work

Let the nominal scan angle (ψ) be zero, indicating that the beam is not intentionally scanned by means of a linear phase progression across the array. As a specific case, we consider a 10-element cos²-FDA whose inter-element distance is $d = \lambda_c/2$, radiating at f_c = 28 GHz, and with Δf = 15 kHz.

We can see the comparison between the Normalized Power Density (NPD in dB) (The NPD is obtained by taking the logarithm with base 10 of the power pattern normalized with respect to the overall maximum radiation value within the considered angle ψ and distance r coverages) obtained with an FDA composed of 10 isotropic elements (with all the parameters set to the above mentioned values) (Figure 2a); and the corresponding NPD of a cos²-FDA presented in this work (Figure 2b), both with $-90° \leq \psi \leq 90°$ and r ranging from 20 to 60 km. Those plots were obtained with the help of Mathematica® software tool [5].

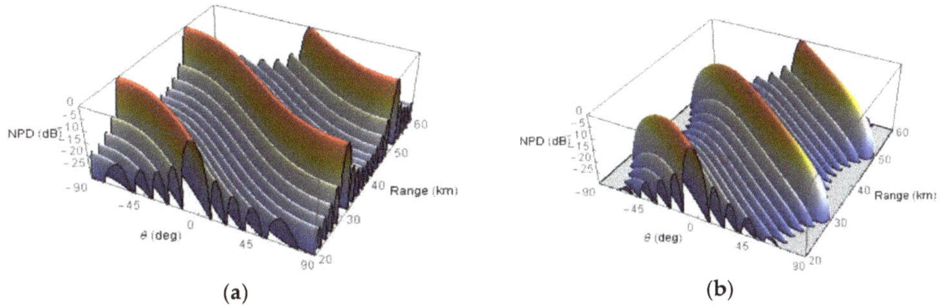

Figure 2. (**a**) Normalized Power Density (NPD) of a 10-element 28-GHz FDA with isotropic elements. (**b**) NPD of a 10-element 28-GHz cos²-FDA.

The beamforming effect for both the isotropic FDA (Figure 3a) and cos²-FDA (Figure 3b) at a given range is obtained by changing the shift frequency. For those plots, a constant range r = 20 km was taken for three values of Δf (10, 15 and 20 kHz).

It can be seen from the above figures that the element power pattern ($\cos^2\psi$ in our case) not only affects, as is well known, the pattern sidelobes (lowering them towards angles deviating from the broadside) but also the steering capabilities of the FDA, thus reducing the main lobe level for both range r (see Figure 2) and frequency shift Δf (see Figure 3) changes. Those effects are not in any way negligible therefore, for practical cases, alternative studies must be performed to correct them.

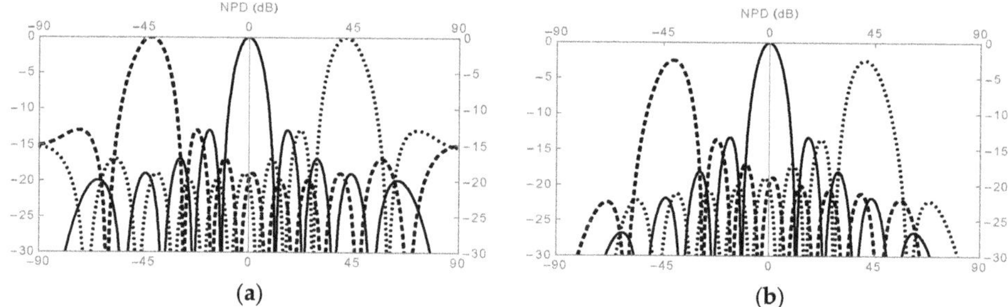

Figure 3. Impact of frequency offset on the main beams. (**a**) Isotropic elements FDA NPD. (**b**) Cos²-FDA NPD. In both cases, r = 20 km, and Δf = 10 kHz: dotted line; Δf = 15 kHz: continuous line; Δf = 20 kHz: dashed line.

The presented numerical simulations show that the use of FDAs at a 5G working frequency (28 GHz) is possible. But from the above considerations, some corrections are needed, even for the general case (i.e., at frequencies not necessarily equal to the one chosen in this work) and this could be taken as a future work. Another research line could be focused on a full-wave simulation of the presented model with a proper electromagnetic simulation software tool.

Author Contributions: Conceptualization, and methodology, M.B.A.A. and J.B.; software, J.B.; Investigation, R.M., J.A.G.-N.; resources, L.C.; writing, M.B.A.A.; Funding acquisition, L.C.

Funding: This research received This work has been funded by the Xunta de Galicia (ED431G2019/01), the Agencia Estatal de Investigación of Spain (TEC2016-75067-C4-1-R, RED2018-102668-T, PID2019-104958RB-C42) and ERDF funds of the EU (FEDER Galicia & AEI/FEDER, UE).

Conflicts of Interest: The authors declare no conflict of interest.

References

1. Kim, T.; Park, J.; Seol, J.; Jeong, S.; Cho, J.; Roh, W. Tens of Gbps Support with Mmwave Beamforming Systems for Next Generation Communications. In Proceedings of the 2013 IEEE Global Communications Conference, Atlanta, GA, USA, 9–13 December 2013; pp. 3679–3685, doi:10.1109/GLOCOM.2013.6831646.
2. Nusenu, S.Y.; Basit, A. Frequency Diverse Array Antennas: From Their Origin to Their Application in Wireless Communication Systems. *J. Comput. Netw. Commun.* **2018**, *2018*, 5815678, doi:10.1155/2018/5815678.
3. Secmen, M.; Demir, S.; Hizal, A.; Eker, T. Frequency Diverse Array Antenna with Periodic Time Modulated Pattern in Range and Angle. In Proceedings of the 2007 IEEE Radar Conference, Boston, MA, USA, 17–20 April 2007; pp. 427–430, doi:10.1109/RADAR.2007.374254..
4. Antonik, P.; Wicks, M. C.; Griffiths, H. D.; Baker, C. J. Frequency Diverse Array Radars, in Proceedings of the IEEE Radar Conference, Verona, WI, USA, 2006; pp. 215–217, doi:10.1109/RADAR.2006.1631800.
5. Hastings, C; Mischo, K.; Morrison, M. *Hands on Start to Wolfram Mathematica*, 2nd ed.; Wolfram Media, Inc.: Champaign, IL, USA, 2016.

Proceedings

Feature Selection in Big Image Datasets †

J. Guzmán Figueira-Domínguez [1,*], Verónica Bolón-Canedo [2] and Beatriz Remeseiro [3]

1 University of A Coruña, 15071 A Coruña, Spain
2 CITIC, University of A Coruña, 15071 A Coruña, Spain; veronica.bolon@udc.es
3 University of Oviedo, 33003 Oviedo, Asturias, Spain; bremeseiro@uniovi.es
* Correspondence: j.guzman.figueira@udc.es

† Presented at the 3rd XoveTIC Conference, A Coruña, Spain, 8–9 October 2020.

Published: 24 August 2020

Abstract: In computer vision, current feature extraction techniques generate high dimensional data. Both convolutional neural networks and traditional approaches like keypoint detectors are used as extractors of high-level features. However, the resulting datasets have grown in the number of features, leading into long training times due to the curse of dimensionality. In this research, some feature selection methods were applied to these image features through big data technologies. Additionally, we analyzed how image resolutions may affect to extracted features and the impact of applying a selection of the most relevant features. Experimental results show that making an important reduction of the extracted features provides classification results similar to those obtained with the full set of features and, in some cases, outperforms the results achieved using broad feature vectors.

Keywords: feature selection; image feature extraction; big data; computer vision

---

1. Introduction

Image datasets have grown not only in the number of samples, but also in the number of features that describe them. At this point, it could be reasonable to expect that having more features would provide more information and better results. However, this does not happen, due to the so-called *curse of dimensionality* [1]. In this context, feature selection [2] contributes to the scalability of the machine learning algorithms by finding the most relevant properties of the images and decreasing train and prediction times. However, their efficiency drastically diminishes when dataset dimension grows. Hence, applying big data technologies may ease to use larger datasets. This article addresses the impact of feature selection on image classification using different feature extraction methods. Particularly, this research focuses on the use of filter methods for feature selection with big data technologies.

2. Materials and Methods

This work proposes a pipeline for image classification composed of three main steps: image feature extraction, feature selection and classification. On the one hand, the first step has been implemented in a *Python* package using Keras, OpenCV and scikit-image libraries. On the other hand, the next steps were developed in an *Apache Spark* application that contains independent jobs for both steps. Additionally, features extracted have been stored in *Kaggle datasets*.

1. Feature extraction: In this work, image feature extraction was performed in order to transform image datasets into columnar feature datasets. The techniques applied here are—*bag of features* methods based on feature detection algorithms like SIFT [3], SURF [4] and KAZE [5]; *linear binary pattern* (LBP) methods [6]; and *convolutional neural networks* (ConvNets) used as feature extractors through architectures like VGG, ResNet and DenseNet.

2. Feature selection: Feature selection includes a broad family of dimensionality reduction techniques that achieve reduction by removing the irrelevant and redundant features while keeping the original relevant ones. Particularly, filter methods select a subset of the original feature set independently of the induction model used. Accordingly, these filter methods are more likely to be applied in a big data scenario due to advantages related to computational costs [7]. In such framework, this research has driven the feature selection stage using the big data platform *Apache Spark* and some implementations of such filter methods: Spark's *MLlib* [8] implementation of the χ^2 filter selector [9]; Spark's implementation of the *Relief-F* method [10]; and *ITFS* framework [11] implementation for Spark [12].
3. Classification: Not every available classifier in *Spark MLlib* has a multi-class nature. So, the suitable models in *Spark* for this problem are *Decision Trees, Random Forests, Naive Bayes* and *Multilayer Perceptron* classifiers. Given the results obtained in the experiments, these two last classifiers were used in the results presented in this manuscript.

In order to carry out the experiments of this research, two datasets were employed—the *ImageNet* dataset, currently hosted by the *Kaggle* platform, which contains 1,281,167 hand-labeled images belonging up to 1000 object categories; and the *Tiny Imagenet* dataset, released as a subset of the original ImageNet, containing very low-resolution images from only a 200-class subset.

3. Results

Regarding results from *Tiny Imagenet*, we noticed that accuracy values provided by features extracted using *bag of features* and *LBP* were quite poor. However, results supplied by features extracted using the ConvNets and applying up to 50% of dimensionality reduction with *Relief-F* (0.6451 top-5 accuracy), χ^2 (0.6422) or *mRMR* (0.6382), outperformed results without feature selection (0.6241).

With respect to experiments carried out with *Imagenet* dataset, features extracted through traditional approaches showed better results with these higher resolution images. Experiments from features extracted using *bag of features*, over the KAZE *keypoints detector*, and applying up to 66% of dimensionality reduction with methods like *mRMR* (0.7674 top-5 accuracy), χ^2 (0.7528) or *ReliefF* (0.7442) showed better results than the ones performed without the selection step (0.7425).

Finally, the accuracy results using features pulled out with a ConvNet like *VGG-19* and feature selection methods were presented quite tight compared to the ones achieved by the own VGG-19 (0.7158 top-1 accuracy and 0.8996 top-5 accuracy). Applying a reduction of a 50% with χ^2 (0.6715 top-1 accuracy and 0.8450 top-5 accuracy) or a reduction of 90% through *mRMR* (0.6554 top-1 accuracy and 0.8143 top-5 accuracy), we notice how results, using a multi-layer perceptron as the classifier model, are below the baseline. However, if we compare the results achieved with a naive Bayes classifier, the baseline (0.6143 top-5 accuracy) is eventually outperformed: 0.6482 top-5 accuracy when applying a reduction up to a 66% with the χ^2 method.

4. Discussion and Conclusions

Contrasting differences on experiments done with all the feature extractors, we can observe some clear tendencies. When feature selection is applied to features extracted with classical techniques, results outperform the baseline collected without making dimensionality reduction. On these techniques, salient information about images is shaped into vectors of a chosen size. As shown in results, this representation may be improved through feature selection techniques. However, when feature selection is applied to *deep features* (i.e., features extracted by pre-trained ConvNets), results are slightly below the baseline without feature selection. This may be explained due to the successive *dropout* layers included in *ConvNets*, which help to remove meaningless information over the layers and represent the best high-order features.

In main terms, results show a clear evidence that feature selection performs a positive impact over features extracted from both datasets. Accuracy values collected in most feature subsets are very close to the ones observed without applying dimensionality reduction. And, in some cases, dimensionality reduction techniques help to outperform classification results using all the features provided by *ConvNets* or *bag of features* extractors. Also, we remark that different feature selection methods stand out depending on the required percentage of feature reduction, so *the best feature selection method* simply does not exist.

Funding: This research has been financially supported in part by European Union FEDER funds, by the Spanish Ministerio de Economía y Competitividad (research project PID2019-109238GB), by the Consellería de Industria of the Xunta de Galicia (research project GRC2014/035), and by the Principado de Asturias Regional Government (research project IDI-2018-000176). CITIC as a Research Centre of the Galician University System is financed by the Consellería de Educación, Universidades e Formación Profesional (Xunta de Galicia) through the ERDF (80%), Operational Programme ERDF Galicia 2014–2020, and the remaining 20% by the Secretaria Xeral de Universidades (ref. ED431G 2019/01).

Acknowledgments: The authors would like to express their gratitude to the CESGA for the provided resources that allowed this research, and the support of NVIDIA Corporation with the donation of Titan Xp GPUs.

Conflicts of Interest: The authors declare no conflict of interest.

References

1. Bellman, R.E. *Dynamic Programming*; Dover Publications, Inc.: New York, NY, USA, 2003.
2. Guyon, I.; Gunn, S.; Nikravesh, M.; Zadeh, L.A. *Feature Extraction: Foundations and Applications*; Springer: Berlin/Heidelberg, Germany, 2008; Volume 207.
3. Lowe, D.G. Distinctive image features from scale-invariant keypoints. *Int. J. Comput. Vis.* **2004**, *60*, 91–110.
4. Bay, H.; Ess, A.; Tuytelaars, T.; Van Gool, L. SURF: Speeded up robust features. *Comput. Vis. Image Underst.* **2008**, *110*, 346–359.
5. Alcantarilla, P.F.; Bartoli, A.; Davison, A.J. KAZE features. In Proceedings of the European Conference on Computer Vision, Firenze, Italy, 7–13 October 2012; pp. 214–227.
6. Ojala, T.; Pietikainen, M.; Harwood, D. A Comparative Study of Texture Measures with Classification Based on Feature Distributions. *Pattern Recognit.* **1996**, *29*, 51–59.
7. Bolón-Canedo, V.; Sánchez-Maroño, N.; Alonso-Betanzos, A. Feature selection for high-dimensional data. *Prog. Artif. Intell.* **2016**, *5*, 65–75.
8. Meng, X.; Bradley, J.; Yavuz, B.; Sparks, E.; Venkataraman, S.; Liu, D.; Freeman, J.; Tsai, D.; Amde, M.; Owen, S.; et al. Mllib: Machine learning in apache spark. *J. Mach. Learn. Res.* **2016**, *17*, 1235–1241.
9. Barnard, G. On the criterion that a given system of deviations from the probable in the case of a correlated system of variables is such that it can be reasonably supposed to have arisen from random sampling. In *Breakthroughs in Statistics*; Springer: Berlin/Heidelberg, Germany, 1992; pp. 1–10.
10. Ramirez, S. RELIEF-F Feature Selection for Apache Spark. Available online: https://github.com/sramirez/spark-RELIEFFC-fselection (accessed on 6 May 2019).
11. Brown, G.; Pocock, A.; Zhao, M.J.; Luján, M. Conditional Likelihood Maximisation: A Unifying Framework for Information Theoretic Feature Selection. *J. Mach. Learn. Res.* **2012**, *13*, 27–66.
12. Ramírez-Gallego, S.; Mouriño-Talín, H.; Martínez-Rego, D.; Bolón-Canedo, V.; Benítez, J.M.; Alonso-Betanzos, A.; Herrera, F. An Information Theory-Based Feature Selection Framework for Big Data Under Apache Spark. *IEEE Trans. Syst. Man Cybern. Syst.* **2018**, *48*, 1441–1453.

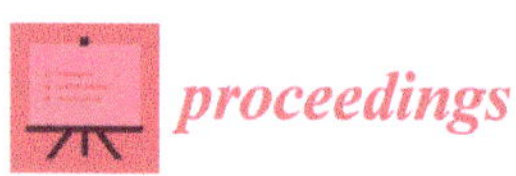

Proceedings

Developing an Open-Source, Low-Cost, Radon Monitoring System †

Alberto Alvarellos * and Juan Ramón Rabuñal

RNASA-IMEDIR, Computer Science Faculty, University of A Coruna, 15071 A Coruña, Spain; juan.rabunal@udc.es
* Correspondence: alberto.alvarellos@udc.es
† Presented at the 3rd XoveTIC Congress, A Coruña, 8–9 October 2020.

Published: 24 August 2020

Abstract: The United States Environmental Protection Agency (USEPA) and the International Agency for Research on Cancer (IARC) have declared Radon gas a human carcinogen. Spain has several regions with high radon concentrations, Galicia (northwestern Spain) being one with the highest Radon concentration. In this work, we present the development of an open-source and low-cost radon monitoring and alert system. The system has two parts: devices and the backend. The devices integrate a Radon sensor, capable of measuring Radon levels every 10 min, and several environmental sensors capable of measuring temperature, humidity, atmospheric pressure, and air pollution. The devices send all the information to the backend, which stores it, exposes it in a web interface, and uses the historical data to predict the radon levels for the following hours. If the radon levels are predicted to overpass the threshold in the next hour, the system issues an alert via several channels (email and MQTT) to the configured recipients for the corresponding device, allowing them to take measures to lower the Radon concentration. The results of this work indicate that the system allows the radon levels to be greatly reduced and makes the development of a low cost and open-source radon monitoring system feasible. The system scalability allows a network of sensors to be created that can help mitigate the health hazard that high radon concentrations create.

Keywords: radon monitoring; IoT; radon alert system; open source; Arduino; Node-RED; open-source

1. Introduction

Radon gas levels are generally high in Spain, and more so in Galicia, caused by its strong links to granite geology, granite being the principal source of radon emissions [1,2]. Several studies carried out in the 1980s highlighted these high radon concentrations [3].

The European Union (EU) has indicated in its guidelines (E2013/59/EURATOM [4]) that annual average radon levels in homes and worksites should not exceed 300 Bq/m^3. These guidelines also indicate the need for member states to include in their Technical Building Codes information regarding radon detection and mitigation. Using a system that can measure radon levels and help mitigate high values, or even better, avoid future high radon levels via predictions, could meet these requirements. In this work, we present the development of the said system. It comprises several monitoring devices and a backend that stores the data, predicts radon levels, and issues alerts.

2. Monitoring Devices

In this work, we design and develop the monitoring devices. They use an RD200M sensor for measuring Radon levels, a BME280 sensor for measuring relative humidity, barometric pressure and ambient temperature, and a CCS811 gas sensor for monitoring indoor air quality. All these data (radon concentration, temperature, humidity, barometric pressure, and air quality) are collected and processed by a processing unit based on the Arduino MKR family (with an ARM Cortex-M0+ CPU). Figure 1a shows the schema of the devices. To integrate all the sensors and the devices' processing unit, we designed an electronic integration board and a 3D printed case (see Figure 1b). The processing unit samples each sensor, processes the data, shows it in a display, and sends it to the backend every 10 min (a sampling period recommended by the radon sensor manufacturer to obtain correct measures). The devices send the data using two different communication technologies: WiFi and Sigfox. We use the WiFi devices in locations where a WiFi network is available, and the Sigfox devices when WiFi is not available.

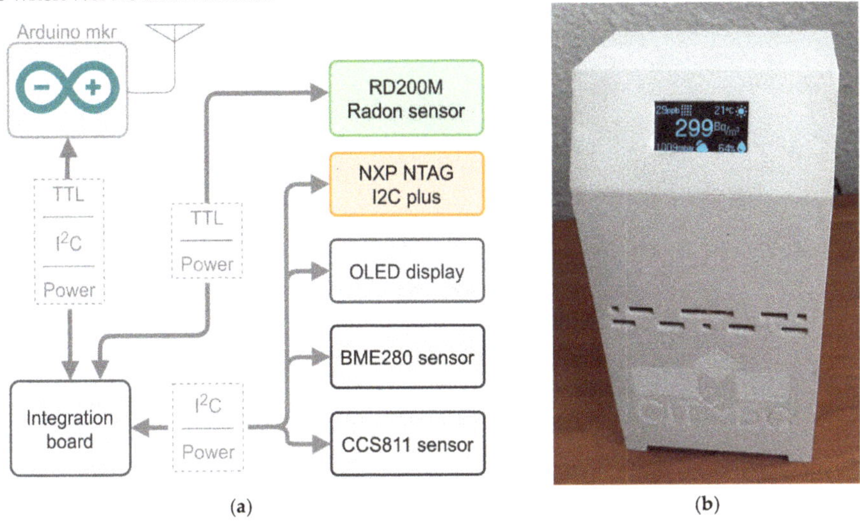

(**a**) (**b**)

Figure 1. (**a**) Monitoring devices sketch. (**b**) Final device assembled in a 3D-printed enclosure.

3. System Architecture

The backend stores the data sent by the devices, it exposes it in a web interface, and predicts the radon level for the next hours, for each device, using its historical data (see Figure 2). If the backend predicts a high radon level for a given device, it sends an MQTT message to an MQTT topic associated with the device, and an email to the addresses registered for the device.

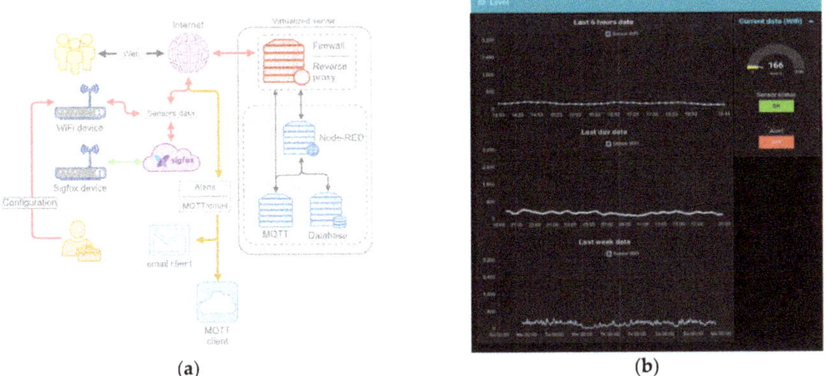

(**a**) (**b**)

Figure 2. (**a**) Architecture of the radon detection and alert system. (**b**) System's web interface.

Once the radon concentration level is below the threshold, the backend sends a message indicating that levels are back to normal (also via MQTT and email). Currently, the system uses the threshold that E2013/59/EURATOM establishes for issuing the alerts (300 Bq/m³), although it can be configured per device.

4. Results

We tested the system in a laboratory with naturally high radon levels in two scenarios. In the first scenario (Figure 3a), the alert system was not activated, showing the natural radon level of the testing site. In the second scenario (Figure 3b), the alert system was activated, and a human operator turned an airflow control system on or off with each alert/"back to normal" message.

Comparing these two scenarios, we can see that the system can satisfactorily be used to significantly lower the radon levels.

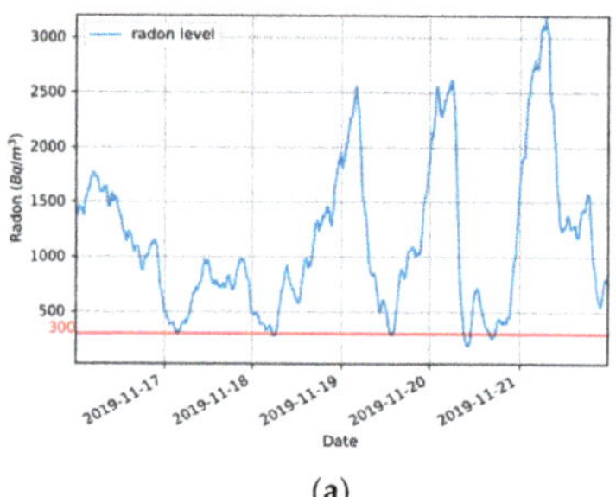

(a)

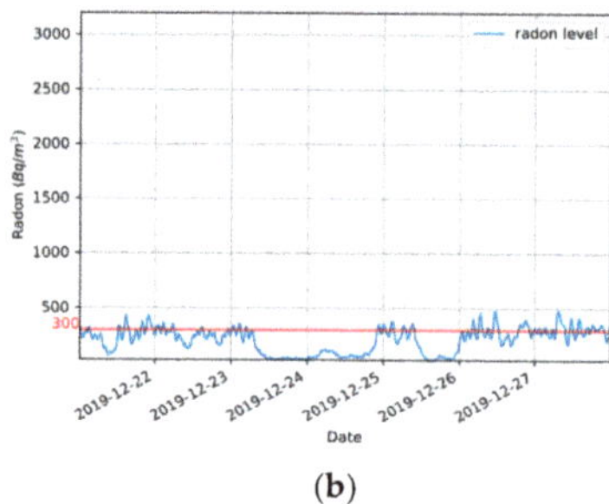

(b)

Figure 3. radon levels for the same location: (**a**) alert system deactivated, (**b**) alert system activated, and airflow control system operated by a human.

Funding: This research received no external funding.

Acknowledgments: we would like to thank the CITEEC's (https://www.udc.es/citeec/) Laboratory of Instrumentation and Intelligent Systems in Civil Engineering staff for their technical support.

Conflicts of Interest: The authors declare no conflict of interest.

References

1. Fleischer, R.L. Radon in the environment—Opportunities and hazards. *Nucl. Tracks Radiat. Meas.* **1988**, *14*, 421–435, doi:10.1016/1359-0189(88)90001-5.
2. Wilkening, M. *Radon in the Environment*; Distributors for the U.S. and Canada, Elsevier Science Pub. Co.: Amsterdam: New York, NY, USA, 1990; ISBN 978-0-444-88163-2.
3. Quindos, L.S.; Fernandez, P.L.; Soto, J. National survey on indoor radon in Spain. *Environ. Int.* **1991**, *17*, 449–453, doi:10.1016/0160-4120(91)90278-X.
4. EUR-Lex Council Directive 2013/59/Euratom of 5 December 2013 Laying down Basic Safety Standards for Protection against the Dangers Arising from Exposure to Ionising Radiation. Available online: https://eur-lex.europa.eu/legal-content/EN/TXT/?uri=CELEX%3A02013L0059-20140117 (accessed on 23 December 2019).

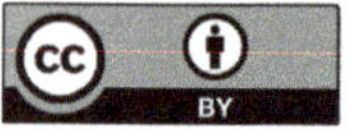

Proceedings

Application of Adaptive Virtual Environments Through Biofeedback for the Treatment of Phobias †

João Donga [1,*], Paulo Veloso Gomes [2], António Marques [2], Javier Pereira [3] and João Azevedo [1]

1 LabRP, Psychosocial Rehabilitation Laboratory, School of Media Arts and Design, Polytechnic Institute of Porto, 4200-465 Porto, Portugal; joaoazevedo@esmad.ipp.pt
2 LabRP, Psychosocial Rehabilitation Laboratory, School of Allied Health Technologies, Polytechnic Institute of Porto, 4200-465 Porto, Portugal; pvg@ess.ipp.pt (P.V.G.); ajmarques@ess.ipp.pt (A.M.)
3 CITIC—Research Center of Information and Communication Technologies, University of A Coruña, 15071 A Coruña, Spain; javier.pereira@udc.es
* Correspondence: jpd@esmad.ipp.pt

† Presented at the 3rd XoveTIC Congress, A Coruña, Spain, 8–9 outubro 2020.

Published: 25 August 2020

Abstract: This study proposes solutions to help people with phobias through the use of virtual environments that allow a contact between the subjects and these phobias. Using neurofeedback, the systems, depending on the emotional state of the user, adapt the scenarios allowing more or less intensity. The phobias these systems treat are social phobia, entomophobia and claustrophobia. The solutions have been developed using Unity, Muse 2 and Vive HTC.

Keywords: social phobia; entomophobia; claustrophobia; virtual reality; immersive environments; brain-computer interface; virtual reality exposure therapy; electroencephalography; biofeedback

---

1. Introduction

A specific phobia consists of fear and anxiety about a particular situation or object. The situation or object is generally avoided when possible; but if exposure does occur, anxiety develops quickly. Anxiety can intensify to the level of a panic attack. People with a specific phobia usually recognize that their fear is irrational and excessive. There are several possible treatments that must be adequate according to the clinical condition. One such treatment is exposure therapy [1]. Patients confront and keep in touch with what they fear and avoid until anxiety gradually decreases through a process called habituation. Typically, therapists start with moderate exposure. When patients are comfortable with an exposure level, the exposure level is increased. Therapists continue to increase the level of the exposure until patients are able to tolerate normal interaction with the situation or object. Using solutions implemented in Virtual Reality (VR) has significative advantages [2], as it is possible to expose patients to situations or objects related to phobias in a controlled manner [3]. In the real world, the use of real animals is very difficult and controlling their behaviors is a complex challenge. The number of animals used also implies costs, and the simulation of situations involves complex logistical issues [4]. The use of a virtual reality-controlled environment allows the therapist to insert the necessary elements to expose the user to the agent that causes the phobia [5]. It is possible to vary its quantity and the intensity of the exposure, and to repeat the process as many times as necessary.

This work intends to reproduce the exposure therapy, and allow that exposure to change in intensity according to the patient's emotional state.

2. Materials and Methods

Two VR solutions were developed with the use of portable electroencephalograms (EEG) that read the user's biofeedback, leading the virtual environments to adapt according to the emotional state via the strategy chosen by the therapist. Virtual Therapy and FearNot are the projects developed by the research group LabRP (Psychosocial Rehabilitation Laboratory) of the School of Allied Health Technologies, and the School of Media Arts and Design, both belonging to the Polytechnic Institute of Porto.

FearNot allows people with insect phobias or claustrophobia to be exposed to environments with insects in varying numbers, or to a room with varying dimensions and brightness, depending on biofeedback reading (Figure 1).

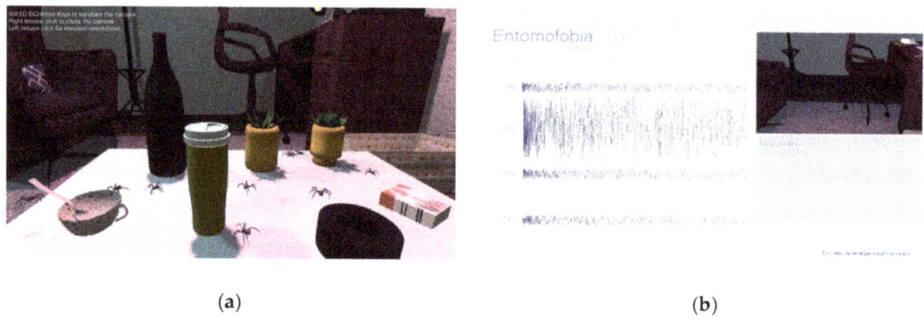

(**a**) (**b**)

Figure 1. (**a**) Virtual environment with spiders; (**b**) Screen used by the therapist to read the electroencephalogram.

Virtual Therapy applies to people with social phobia, and exposes the patient to a social experience in a professional context that consists of an oral presentation to the public in an auditorium. It designs an automatic response to user performance based on biomedical data, and varies the number of spectators in the auditorium and their behavior (Figure 2).

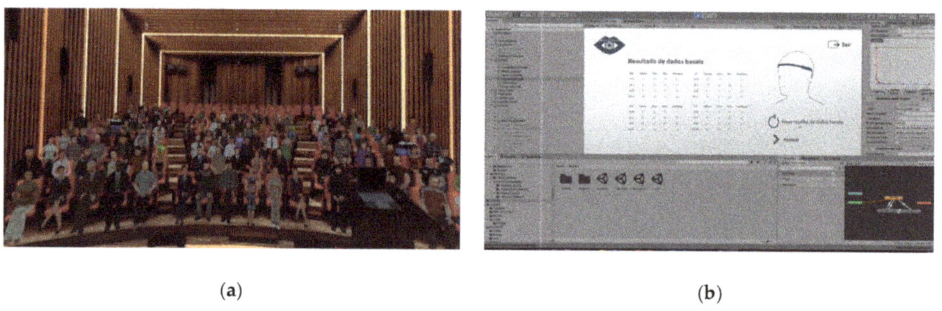

(**a**) (**b**)

Figure 2. (**a**) Virtual environment simulating an auditorium; (**b**) Control Menu.

3. Results

For the time being, no results are available to draw definitive conclusions about the impact of this study. Even so, the observations made during the tests reflects the fact that the participants show a high level of motivation, interest and involvement towards the activities developed. However, it is necessary to wait for the final collection of data and its subsequent analysis, as well as future tests with larger samples, to determine if the project has a real impact on the quality of life of people with the related phobias.

Author Contributions: Conceptualization, J.D.; methodology, J.D. and P.V.G.; writing—review and editing, J.D.; P.V.G. and J.A.; supervision, A.M. and J.P; project administration, A.M. and J.P. All authors have read and agreed to the published version of the manuscript.

Funding: This research received no external funding.

Acknowledgments: The authors would like to thank the Multimedia students that developed Virtual Therapy (Catarina Baldaia, Gil Araújo) and FearNot (Daniela Martins, Mariana Carvalho, Margarida Santos).

Conflicts of Interest: The authors declare no conflict of interest.

References

1. Gorini, A.; Riva, G. Virtual reality in anxiety disorders: the past and the future; Expert Review of Neurotherapeutics, **2008**, *8*, 215–233. doi:10.1586/14737175.8.2.215
2. Cláudio, A.P.; Carmo, M.B.; Pinheiro, T.; Esteves, F.; Lopes, E. Virtual Environment to Treat Social Anxiety; In *BT—Design, User Experience, and Usability. Health, Learning, Playing, Cultural, and Cross-Cultural User Experience*; Marcus, A., Ed.; Springer: Berlin/Heidelberg, Germany, 2013; pp. 442–451.
3. Gebara, C.M.; Barros-Neto, T.P.; Gertsenchtein, L.; Lotufo-Neto, F. Virtual reality exposure using three-dimensional images for the treatment of social phobia. *Braz. J. Psychiatry* **2016**, *38*, 24–29. doi:10.1590/1516-4446-2014-1560
4. Musalek, M.; Vasek, L. Possibilities of using virtual reality as a means for therapy from fear of spiders. In *MATEC Web of Conferences;* EDP Sciences: Ulis, France, 2019; Volume 292.
5. Meyerbröker, K.; Emmelkamp, P.M.G. Virtual Reality Exposure Therapy for Anxiety Disorders: The State of the Art. In *Advanced Computational Intelligence Paradigms in Healthcare 6*; Springer: Berlin/Heidelberg, Germany, 2011; Volume 337, pp. 47–62.

Proceedings

The Use of Portable EEG Devices in Development of Immersive Virtual Reality Environments for Converting Emotional States into Specific Commands †

Catarina Sá *, Paulo Veloso Gomes, António Marques and António Correia

LabRP, Psychosocial Rehabilitation Laboratory, School of Allied Health Technologies, Polytechnic Institute of Porto, 4200-072, Porto, Portugal; pvg@ess.ipp.pt (P.V.G.); ajmarques@ess.ipp.pt (A.M.); amsc@ess.ipp.pt (A.C.)

* Correspondence: cfds@ess.ipp.pt

† Presented at the 3rd XoveTIC Congress, A Coruña, Spain, 8–9 October 2020.

Published: 25 August 2020

Abstract: The application of electroencephalography electrodes in Virtual Reality (VR) glasses allows users to relate cognitive, emotional, and social functions with the exposure to certain stimuli. The development of non-invasive portable devices, coupled with VR, allows for the collection of electroencephalographic data. One of the devices that embraced this new trend is Looxid Link™, a system that adds electroencephalography to HTC VIVE™, VIVE Pro™, VIVE Pro Eye™, or Oculus Rift S™ glasses to create interactive environments using brain signals. This work analyzes the possibility of using the Looxid Link™ device to perceive, evaluate and monitor the emotions of users exposed to VR.

Keywords: mental health and wellness; emotions; empathy; immersive environments; virtual reality; electroencephalography; neurogaming; neurofeedback

---

1. Introduction

In the last few years, virtual reality (VR) has become more popular [1]. VR is a technology that allows users to generate, through a computer, three-dimensional (3D) virtual worlds identical to the real world. It can create immersive experiences, in real time, and it is possible for the user to interact with the environment as if it is the real world [2–4]. Since the 1970s, VR has been used in several applications, such as in industry, education, medicine, and scientific research. Since then VR has been used in developed tests, training and treatment approaches [5,6].

Technological development has allowed VR devices, in addition to having better image and sound quality, to have reduced size and weight. These characteristics allow for its use to provide greater mobility on the part of the user, adding the movement of the human body to the exploration of the surrounding space [5,6].

The technological evolution of VR equipment and the creation of immersive environments that simulate real environments and situations, open doors to its applications in more complex areas, such as neurogaming and neurofeedback therapies. Brain activity studies obtained by the application of electroencephalography (EEG) electrodes have allowed users to relate cognitive, emotional, and social functions with the exposure to certain stimuli. The portability and mobility that current VR devices allow are incompatible with traditional EEG equipment, require user immobilization and a panoply of wires and equipment that limit their use and interfere with the user experience [7–9].

In order to make the collection of EEG data less invasive, and to ensure that this collection is made without the user being aware, portable devices have been developed, which can be attached to Virtual Reality glasses [7,8]. One of the devices that embraced this new trend is the Lookxid Link™, which, in addition to having all the features offered by a regular VR headset, also has an EEG attached. This device detects EEG signals and uses an API that converts electroencephalographic patterns into commands that can be applied to the VR environment [10]. The EEG in the VR headset can bring several advantages, such as following the cognitive state of the user and measuring the user's brain activity [7,8].

Lookxid Link™ is a system that adds EEG channels to HTC VIVE, VIVE Pro or Oculus Rift S, to create interactive environments using brain signals. In this equipment, the sensors that are used are AF3, AF4, AF7, AF8, Fp1 and Fp2 of the International System 10-10, located in the Pre-Frontal Cortex [10].

The Pre-Frontal Cortex (PFC) is the anterior area of the frontal cerebral lobe that is divided into the primary motor cortex and the pre-motor cortex that are located posterior to the PFC [11]. The PFC is responsible for executive functions (planning, decision making, inhibitory control/weighting, attention and working memory—ability to retain information for the execution of an action), social behavior, emotions, affection for others and intelligence, as well as other cognitive controls [11–13].

Emotions appear as a response to the perception of an object or situation, and human beings communicate more easily through emotional expressions [14,15]. Emotions have been studied to understand the interpretation and processing of emotions at the cortical level. To recognize emotions through EEG signals, it is necessary to pay attention to several aspects, such as the characteristics of the time domain, the frequency domain and the time frequency of the EEG signals, in order to have a correlation of information between the different channels of EEG [14]. Negative emotions are believed to be closely related to the right hemisphere, while positive emotions are processed by the left hemisphere [16,17].

2. Objectives

This work analyzes the possibility of using the Looxid Link™ device to perceive, evaluate and monitor the emotions of users exposed to immersive environments in VR, converting the signals captured by the different EEG channels into commands that influence the stimuli generated by the system, according to the emotional state of the user.

3. Methods

This article is a bibliographic review regarding the use of portable EEG devices in the development of immersive virtual reality environments for converting emotional states into specific commands. The information was collected using research platforms, namely PubMed—NCBI, the EBSCO Information Services, B-on and Google Academic.

For this research, the terms used were virtual reality, electroencephalography, Lookxid Link™, neurofeedback, emotions and neurogaming.

However, no articles were found in which Lookxid Link™ was used as an object of investigation.

4. Discussion and Conclusions

Lookxid Link™, despite having only the prefrontal electrodes, as does the EEG coupling in the VR headset, the mobility and artifact problems no longer exist, these being the great advantages of using this equipment instead of two separate units [10].

With the use of Looxid Link™, it is possible to monitor and control the electroencephalographic signals and consequently the emotions that the user may be feeling [10]. The individual's relaxation provides an increase in the theta rhythm in all EEG channels, while if the user is attentive it will lead to an increase in beta in all channels [18]. If the user is afraid, there will be an increase in beta waves on the left side [19].

Author Contributions: Conceptualization, C.S. and P.V.G.; methodology, C.S., P.V.G. and A.M.; validation, P.V.G., A.M. and A.C.; investigation, C.S.; writing—original draft preparation, C.S.; writing—review and editing, C.S. and P.V.G.; visualization, C.S., P.V.G and A.C.; supervision, P.V.G. and A.M.; project administration, C.S.

Funding: This research received no external funding.

Acknowledgments: This research was carried out and used the equipment of the Psychosocial Rehabilitation Laboratory (LabRp) of the Research Center in Rehabilitation of the School of Allied Health Technologies, Polytechnic Institute of Porto.

Conflicts of Interest: The authors declare no conflict of interest.

References

1. O'Connor, S. Virtual Reality and Avatars in Health care. In *Clinical Nursing Research*; SAGE Publications Inc.: Thousand Oaks, CA, USA, 2019; Volume 28, pp. 523–528.
2. Freeman, D.; Reeve, S.; Robinson, A.; Ehlers, A.; Clark, D.; Spanlang, B.; Slater, M. Virtual reality in the assessment, understanding, and treatment of mental health disorders. *Psychol. Med.* **2017**, *47*, 2393–2400.
3. Pandrangi, V.C.; Gaston, B.; Appelbaum, N.P.; Albuquerque, F.C.; Levy, M.M.; Larson, R.A. The Application of Virtual Reality in Patient Education. *Ann. Vasc. Surg.* **2019**, *59*, 184–189.
4. Botella, C.; Fernández-Álvarez, J.; Guillén, V.; García-Palacios, A.; Baños, R. Recent Progress in Virtual Reality Exposure Therapy for Phobias: A Systematic Review. *Curr. Psychiatry Rep.* **2017**, *19*, 42.
5. El Beheiry, M.; Doutreligne, S.; Caporal, C.; Ostertag, C.; Dahan, M.; Masson, J. Virtual Reality: Beyond Visualization. *J. Mol. Biol.* **2019**, *431*, 1315–1321.
6. Rizzo, A.; Koenig, S.T. Is Clinical Virtual Reality Ready for Primetime? *Association* **2017**, *31*, 877–899.
7. Tremmel, C.; Herff, C.; Krusienski, D.J. EEG Movement Artifact Suppression in Interactive Virtual Reality. In Proceedings of the Annual International Conference of the IEEE Engineering in Medicine and Biology Society, EMBS, Berlin, Germany, 23–27 July 2019; pp. 4576–4579.
8. Tauscher, J.P.; Schottky, F.W.; Grogorick, S.; Bittner, P.M.; Mustafa, M.; Magnor, M. Immersive EEG: Evaluating electroencephalography in virtual reality. In Proceedings of the 26th IEEE Conference on Virtual Reality and 3D User Interfaces, VR 2019, Osaka, Japan, 23–27 March 2019; pp. 1794–1800.
9. Torner, J.; Skouras, S.; Molinuevo, J.L.; Gispert, J.D.; Alpiste, F. Multipurpose Virtual Reality Environment for Biomedical and Health Applications. *IEEE Trans. Neural Syst. Rehabil. Eng.* **2019**, *27*, 1511–1520.
10. Lookxid Labs. Lookxid Link. 2020. [Online]. Available online: https://looxidlabs.com/looxidlink/ (accessed on 20 July 2020).
11. Joaquín, F.M. *The Prefrontal Cortex*, 4th ed.; Elsevier Ltd.: San Diego, CA, USA, 2008.
12. Ko, J. Neuroanatomical Substrates of Rodent Social Behavior: The Medial Prefrontal Cortex and Its Projection Patterns. *Front. Neural Circuits* **2017**, *11*, 41.
13. Fuster, J.M. The prefrontal cortex—An update: Time is of the essence. *Neuron* **2001**, *30*, 319–333.
14. Chao, H.; Zhi, H.; Dong, L.; Liu, Y. Recognition of Emotions Using Multichannel EEG Data and DBN-GC-Based Ensemble Deep Learning Framework. *Comput. Intell. Neurosci.* **2018**, *2018*, 11.
15. Zhao, G.; Zhang, Y.; Ge, Y. Frontal EEG asymmetry and middle line power difference in discrete emotions. *Frontiers* **2018**, *12*, 225.
16. Bekkedal, M.Y.V.; Rossi, J.; Panksepp, J. Neuroscience and Biobehavioral Reviews Human brain EEG indices of emotions: Delineating responses to affective vocalizations by measuring frontal theta event-related synchronization. *Neurosci. Biobehav. Rev.* **2011**, *35*, 1959–1970.
17. Id, A.Y.S.; Kasinathan, K.; Karapoondinott, V. Individual EEG measures of attention, memory, and motivation predict population level TV viewership and Twitter engagement. *PLoS ONE* **2019**, *14*, e0214507.
18. Liu, N.H.; Chiang, C.Y.; Chu, H.C. Recognizing the degree of human attention using EEG signals from mobile sensors. *Sensors* **2013**, *13*, 10273–10286.
19. Bălan, O.; Moise, G.; Moldoveanu, A.; Leordeanu, M.; Moldoveanu, F. An investigation of various machine and deep learning techniques applied in automatic fear level detection and acrophobia virtual therapy. *Sensors* **2020**, *20*, 496.

Proceedings

Enhancing Retinal Blood Vessel Segmentation through Self-Supervised Pre-Training †

José Morano [1,2,*], Álvaro S. Hervella [1,2], Noelia Barreira [1,2], Jorge Novo [1,2] and José Rouco [1,2]

1 CITIC-Research Center of Information and Communication Technologies, University of A Coruña, 15071 A Coruña, Spain; a.suarezh@udc.es (Á.S.H.); noelia.barreira@udc.es (N.B.); jnovo@udc.es (J.N.); jrouco@udc.es (J.R.)

2 Department of Computer Science, University of A Coruña, 15071 A Coruña, Spain

* Correspondence: j.morano@udc.es

† Presented at the 3rd XoveTIC Conference, A Coruña, Spain, 8–9 October 2020.

Published: 25 August 2020

Abstract: The segmentation of the retinal vasculature is fundamental in the study of many diseases. However, its manual completion is problematic, which motivates the research on automatic methods. Nowadays, these methods usually employ Fully Convolutional Networks (FCNs), whose success is highly conditioned by the network architecture and the availability of many annotated data, something infrequent in medicine. In this work, we present a novel application of self-supervised multimodal pre-training to enhance the retinal vasculature segmentation. The experiments with diverse FCN architectures demonstrate that, independently of the architecture, this pre-training allows one to overcome annotated data scarcity and leads to significantly better results with less training on the target task.

Keywords: self-supervised learning; transfer learning; multimodal; retinal vasculature segmentation

1. Introduction

Retinal vasculature segmentation represents a key step in the analysis of multiple common diseases like glaucoma and diabetes. However, its manual completion is arduous and partly subjective, so automatic methods have emerged as an advantageous alternative. State-of-the-art vasculature segmentation is based on Fully Convolutional Networks (FCNs). Nonetheless, using FCNs requires addressing two major difficulties: (1) Determining the network architecture and (2) gathering a large amount of annotated training data. The first issue can be partly overcome by reviewing similar problems. Annotated data, however, are usually scarce in medical imaging, as they require experts to be involved in a tedious process. This motivates the proposal of self-supervised multimodal pre-training (SSMP) to learn the relevant patterns from unlabeled data and reduce the required amount of annotated data [1–3]. Specifically, the proposed SSMP consists of training an FCN to predict fluorescein angiographies (a grayscale modality that enhances the vasculature) from retinographies.

In this work, we present a novel application of SSMP to enhance vasculature segmentation in a transfer learning setting, performing a comparative analysis of several FCN architectures.

2. Methodology

The main objective of this work is the segmentation of the retinal vasculature using FCNs. To enhance the results, we propose a transfer learning setting that consists of using SSMP followed by a fine-tuning in the segmentation task [4]. To appraise our proposal, we evaluated the results of the same networks using the SSMP or training from scratch and with different training set sizes (1, 5, 10, and 15). In all of the cases, we used the following FCN architectures: U-Net [5], FC-DenseNet [6], and ENet [7,8].

In order to perform the SSMP, we aligned the 59 retinography–angiography pairs of the publicly available Isfahan MISP dataset [9] using the method proposed in [10]. Then, inspired by [1,2], we used SSIM function to compute the reconstruction loss between the network output and its ground truth.

To train the networks for the vasculature segmentation task, we employed the DRIVE dataset [11], which consists of 40 retinographies and their corresponding vasculature segmentation masks. As the loss, we used Binary Cross-Entropy. For testing, we included the 20 annotated images of the STARE dataset [12].

The networks were trained using the Adam optimization algorithm with learning rate decay and data augmentation through affine transformations and color and intensity variations.

3. Results and Conclusions

Table 1 shows the best AUC-ROC and AUC-PR values of the different networks trained from scratch (FS) and using SSMP for the STARE dataset. Moreover, in Figure 1 is depicted an example of the segmentation masks predicted by the U-Net trained with 15 images, with and without SSMP. As observed, the use of SSMP has significant benefits in both quantitative and qualitative terms; mainly due to the fact that the vessel continuity is better preserved and the pathological structures are better handled. This improvement, in addition, is achieved with less training in the target task. These results demonstrate that the use of SSMP emerges as a valuable option when annotated data in the target task are scarce.

Regarding the diverse FCN architectures, both qualitative and quantitative results (see Table 1) demonstrated that the U-Net provided the best performance.

Table 1. Best AUC-ROC and AUC-PR values of the different networks trained from scratch (FS) and using self-supervised multimodal pretraining (SSMP) for the STARE dataset.

U-Net				FC-DenseNet				ENet			
SSMP		FS		SSMP		FS		SSMP		FS	
ROC	PR	ROC	PR	ROC	PR	ROC	PR	ROC	PR	ROC	PR
0.9834	0.9051	0.9728	0.8590	0.9794	0.8924	0.9699	0.8468	0.9694	0.8434	0.8349	0.4472

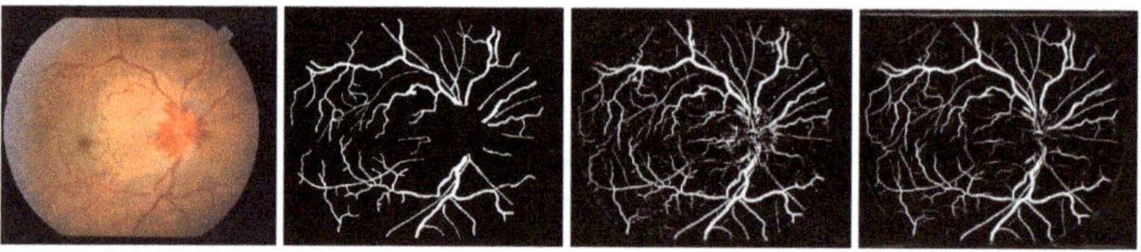

Figure 1. Example of the predicted vasculature mask. From left to right: Original STARE retinography, vasculature segmentation ground truth, vasculature segmentation mask predicted by U-Net trained with 15 images without SSMP, vasculature mask predicted by the same network with SSMP.

Author Contributions: Conceptualization, Á.S.H., J.N., and J.R.; methodology, Á.S.H. and J.M.; software, Á.S.H. and J.M.; validation, Á.S.H. and J.M.; formal analysis, Á.S.H. and J.M.; investigation, J.M.; resources, N.B., J.N., and J.R.; data curation, Á.S.H. and J.M.; writing—original draft preparation, J.M.; writing—review and editing, J.M. and J.N.; visualization, J.M.; supervision, J.N. and J.R.; project administration, J.N. and J.R.; funding acquisition, J.N. and J.R. All authors have read and agreed to the published version of the manuscript.

Funding: This work is supported by the Instituto de Salud Carlos III, Government of Spain, and FEDER funds of the European Union through the DTS18/00136 research projects and by the Ministerio de Ciencia, Innovación y Universidades, Government of Spain through the RTI2018-095894-B-I00 research projects. In addition, this work has received financial support from the Consellería de Educación, Universidade e Formación Profesional, Xunta de Galicia, through the ERDF (80%) and Secretaría Xeral de Universidades (20%), CITIC, Centro de Investigación del Sistema Universitario de Galicia, Ref. ED431G 2019/01.

Conflicts of Interest: The authors declare no conflict of interest. The founding sponsors had no role in the design of the study; in the collection, analyses, or interpretation of data; in the writing of the manuscript, and in the decision to publish the results.

References

1. Hervella, Á.S.; Rouco, J.; Novo, J.; Ortega, M. Retinal Image Understanding Emerges from Self-Supervised Multimodal Reconstruction. In Proceedings of the Medical Image Computing and Computer Assisted Intervention—MICCAI, Granada, Spain, 16–20 September 2018; Volume 11070, pp. 321–328.
2. Álvaro S. Hervella.; Rouco, J.; Novo, J.; Ortega, M. Self-supervised multimodal reconstruction of retinal images over paired datasets. *Expert Syst. Appl.* **2020**, *161*, 113674.
3. Hervella, Á.S.; Rouco, J.; Novo, J.; Ortega, M. Learning the retinal anatomy from scarce annotated data using self-supervised multimodal reconstruction. *Appl. Soft Comput.* **2020**, *91*, 106210. doi:10.1016/j.asoc.2020.106210.
4. Morano, J.; Hervella, Á.S.; Barreira, N.; Novo, J.; Rouco, J. Multimodal Transfer Learning-based Approaches for Retinal Vascular Segmentation. In Proceedings of the European Conference on Artificial Intelligence (ECAI), Santiago de Compostela, Spain, 29 August–5 September 2020.
5. Ronneberger, O.; P.Fischer.; Brox, T. U-Net: Convolutional Networks for Biomedical Image Segmentation. In Proceedings of the Medical Image Computing and Computer-Assisted Intervention (MICCAI), Munich, Germany, 5–9 October 2015; Volume 9351, pp. 234–241.
6. Jégou, S.; Drozdzal, M.; Vazquez, D.; Romero, A.; Bengio, Y. The One Hundred Layers Tiramisu: Fully Convolutional DenseNets for Semantic Segmentation. In Proceedings of the 2017 IEEE Conference on Computer Vision and Pattern Recognition Workshops (CVPRW), Honolulu, HI, USA, 21–26 July 2017; pp. 1175–1183. doi:10.1109/CVPRW.2017.156.
7. Paszke, A.; Chaurasia, A.; Kim, S.; Culurciello, E. ENet: A Deep Neural Network Architecture for Real-Time Semantic Segmentation. *CoRR* **2016**, *abs/1606.02147*.
8. Canziani, A.; Culurciello, E.; Paszke, A. Evaluation of neural network architectures for embedded systems. In Proceedings of the 2017 IEEE International Symposium on Circuits and Systems (ISCAS), Baltimore, MD, USA, 28–31 May 2017; pp. 1–4. doi:10.1109/ISCAS.2017.8050276.
9. Kashefpur, M.; Kafieh, R.; Jorjandi, S.; Golmohammadi, H.; Khodabande, Z.; Abbasi, M.; Fakharzadeh, A.A.; Kashefpoor, M.; Rabbani, H. Isfahan MISP Dataset. *J. Med. Signals Sens.* **2016**, *7*, 43–48.
10. Hervella, Á.S.; Rouco, J.; Novo, J.; Ortega, M. Multimodal registration of retinal images using domain-specific landmarks and vessel enhancement. *Procedia Comput. Sci.* **2018**, *126*, 97 – 104.
11. Staal, J.; Abramoff, M.D.; Niemeijer, M.; Viergever, M.A.; van Ginneken, B. Ridge-based vessel segmentation in color images of the retina. *IEEE Trans. Med. Imaging* **2004**, *23*, 501–509. doi:10.1109/TMI.2004.825627.
12. Hoover, A.D.; Kouznetsova, V.; Goldbaum, M. Locating blood vessels in retinal images by piecewise threshold probing of a matched filter response. *IEEE Trans. Med. Imaging* **2000**, *19*, 203–210. doi:10.1109/42.845178.

Proceedings

Study on Relevant Features in COVID-19 PCR Tests †

Plácido L. Vidal [1,2,*], Joaquim de Moura [1,2], Lucía Ramos [1,2], Jorge Novo [1,2] and Marcos Ortega [1,2]

1 Centro de investigación CITIC, Campus de Elviña, Universidade da Coruña, s/n, 15071 A Coruña, Spain; joaquim.demoura@udc.es (J.d.M.); l.ramos@udc.es (L.R.); jnovo@udc.es (J.N.); mortega@udc.es (M.O.)
2 Grupo VARPA, Instituto de Investigación Biomédica de A Coruña (INIBIC), Universidade da Coruña, Xubias de Arriba, 84, 15006 A Coruña, Spain
* Correspondence: placido.francisco.lizancos.vidal@udc.es; Tel.: +34-981-16-70-00 (ext. 1330)
† Presented at the 3rd XoveTIC Conference, A Coruña, Spain, 8–9 October 2020.

Published: 26 August 2020

Abstract: In the year 2020, the world suffered the effects of a global pandemic. COVID-19 is a disease that mainly affects the respiratory system of patients, even causing a disproportionate response of the immune system and further spreading the damage to other vital organs. The main means by which health care services detected this viral disease was through the use of Polymerase Chain Reactions (PCRs). These PCRs allow the detection of known chains of the genetic code of the virus in samples of sputum. In this work, we study PCR signal features that allow to automatize the analysis of hundreds of PCRs. The findings obtained from the study have shown these features to be capable of obtaining successful results in the detection of COVID-19 in PCR samples, with only a small fraction of the information extracted by the clinicians for that purpose.

Keywords: polymerase chain reaction; COVID-19; feature analysis

---

1. Introduction

SARS-CoV-2 is a strain of coronavirus responsible for the global COVID-19 pandemic of 2020. This virus causes a severe acute respiratory syndrome (SARS-CoV-2) and viral pneumonia that may leave lungs severely and irreparably damaged [1,2].

For the detection of this virus, sputum samples are analyzed. In these tests, using RT-qPCR or Reverse Transcription Quantitative Polymerase Chain Reaction, the RNA chains that are present in the obtained samples are transcribed into DNA. These tests use deoxyribonucleotides with fluorescent markers that, as soon as they successfully couple with the reference sequence, emit light. This way, by measuring the fluorescence emitted during each cycle, we can know if the reference gene has been found. The gene to be detected is called "E", common to the family to which COVID-19 belongs to group 2 coronavirus.

2. Methodology

To study these signals, we will analyze different features, in the search for one that best separates positive from negative patients. In this case, we have analyzed the mean, standard deviation and the percentiles 25% and 75% of the PCR signals of the patient (as well as the positive and negative reference signals for the PCR batch of that given sample). To study their separability, we will use a kernel density estimation with a gaussian kernel over a trained support vector machine (SVM) with a lineal kernel.

3. Results

Our dataset is composed by 65 positive and 65 negative patients. These signals have been generated with a LightCycler 480 Real-Time PCR System from Roche, using fluorophores with wavelength of excitation of 465 nm and wavelength of detection of 510 nm over 45 PCR cycles. In Figure 1, we can see the mean relative fluorescence returned by the positive PCRs and the negative PCRs. This figure clearly shows the general condition that separates a positive sample from a negative one: the excitation of the signal that is obtained after a certain number of PCR cycles. However, this slope and cycle threshold is highly dependent on the batch (thus the need for the reference signals).

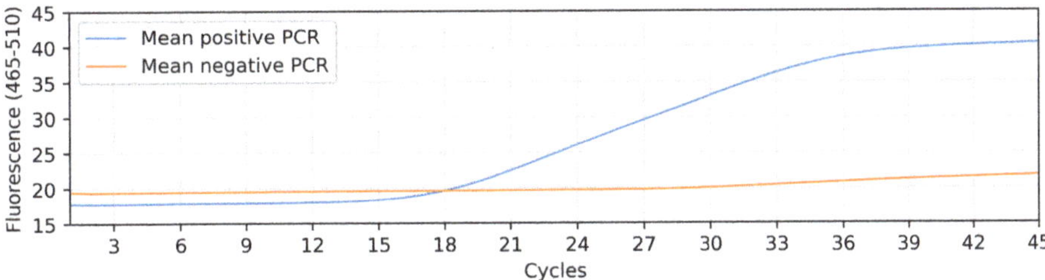

Figure 1. Mean PCR fluorescence signals per cycle of all positive and negative patients of the dataset.

Nonetheless, the best results were obtained by using only the standard deviation of the main signal, with an F1 score of 0.94. As shown in Figure 2, we can satisfactorily classify the majority of the patients without the necessity of the reference signals of the PCRs and with a minimal overlap between classes.

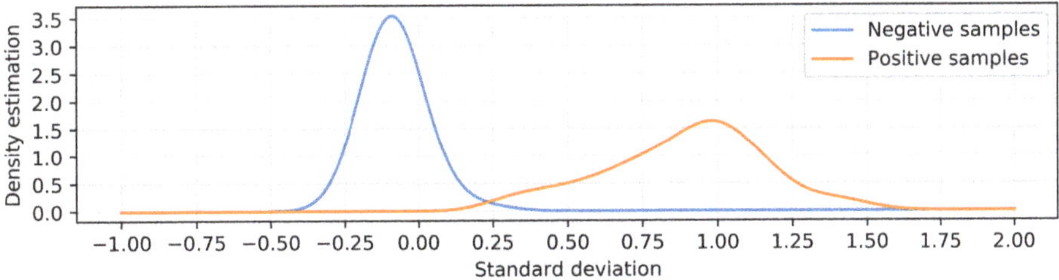

Figure 2. Resulting density estimation for the trained classifier using a kernel bandwith of 0.1.

As future work, we plan a more in-depth study with a larger dataset. This will allow us to generate more complex features, as well as using modern machine learning and statistical strategies to develop a robust system that can, effectively, further speed up the process of diagnosing this disease.

Author Contributions: All the authors contributed equally to this work and have read and agreed to the published version of the manuscript.

Funding: Instituto de Salud Carlos III, Government of Spain, DTS18/00136 research project; Ministerio de Ciencia, Innovación y Universidades, Government of Spain, RTI2018-095894-B-I00 research project, Ayudas para la formación de profesorado universitario (FPU), grant ref. FPU18/02271; CITIC, Centro de Investigación de Galicia ref. ED431G 2019/01, receives financial support from Consellería de Educación, Universidade e Formación Profesional, Xunta de Galicia, through the ERDF (80%) and Secretaría Xeral de Universidades (20%).

Conflicts of Interest: The authors declare no conflict of interest. The funders had no role in the design of the study; in the collection, analyses, or interpretation of data; in the writing of the manuscript, or in the decision to publish the results.

References

1. De Moura, J.; Novo, J.; Ortega, M. Fully automatic deep convolutional approaches for the analysis of Covid-19 using chest X-ray images. *medRxiv* **2020**, doi:10.1101/2020.05.01.20087254.
2. De Moura, J.; Ramos, L.; Vidal, P. L.; Cruz, M.; Abelairas, L.; Castro, E.; Novo, J.; Ortega, M. Deep convolutional approaches for the analysis of Covid-19 using chest X-Ray images from portable devices. *medRxiv* **2020**, doi:10.1101/2020.06.18.20134593.

Proceedings

Mobile Application for Analysing the Development of Motor Skills in Children †

David Moreno Naya [1], Francisco J. Vazquez-Araujo [2], Paula M. Castro [2], Jamile Vivas Costa [1], Adriana Dapena [2,*] and Luz González Doniz [1]

1 Research Group Intervención Psicosocial y Rehabilitación Funcional, Department of Physiotherapy, Medicine and Biomedical Sciences, University College of Physiotherapy, University of A Coruña, Campus de Oza, 15006 A Coruña, Spain; david.moreno@udc.es (D.M.N.); j.vivas@udc.es (J.V.C.); luz.doniz@udc.es (L.G.D.)

2 Research Group Tecnología Electrónica y Comunicaciones, Department of Computer Engineering, Faculty of Computer Science, CITIC Research Center & University of A Coruña, Campus de Elviña, 15071 A Coruña, Spain; francisco.vazquez@udc.es (F.J.V.-A.); paula.castro@udc.es (P.M.C.)

* Correspondence: adriana.dapena@udc.es

† Presented at the 3rd XoveTIC Conference, A Coruña, Spain, 8–9 October 2020.

Published: 26 August 2020

Abstract: This work presents a mobile application to complement and reinforce the specific physical activities in children through training prior to such activities and monitoring their progress after it. This experiment has been developed on a healthy population of children from an education centre in the area of A Coruña. The results show increasing errors for lower primary school years, as expected, and also strongly dependent on the motor path type or characteristic. Therefore, this tool will be suitable for use with children affected by motor coordination difficulties.

Keywords: data science; engine development; information systems; mobile app; motor coordination

---

1. Introduction

A child's motor development involves a sequential evolution from initial simple and disorganized movements to increasingly complex and organized movements [1]. However, in the stages of infant and primary schooling (3–6 years and 7–12 years, respectively), it is common to detect the need for educational reinforcement to improve the learning of these motor skills. Appropriate early intervention could correct some of these difficulties and contribute to the child's motor progress thus avoiding possible negative implications such as poor motor coordination [2]. This work presents a mobile application, referred to as simply app, for the analysis of the motor evolution in these children. This tool is useful for any professional directly involved with children's motor development, such as psychopedagogues, physiotherapists, occupational therapists and, of course, the children themselves [3]. The tool developed consists of the tactile tracking on the screen of the mobile device of an object that moves on different paths designed by an interdisciplinary team, made up of physiotherapists and computer specialists. The path characteristics and the object movement can be configured by the expert according to the initial assessment of each child and the periodic feedback received from the analysis of the data extracted using this application throughout different sessions. The app offers an objective information system that is indispensable for quantifying the child's progress through an adequate data registration, data export and a subsequent analysis that allows a detailed follow-up of the detected need. Moreover, the tool can be adapted to this evolution and to each individual need for reinforcement. Thus, aspects such as the path to be followed, the characteristics of this path, the type of object that moves on it, the speed factor of this object when moving on this path or the number of attempts, among others, can be configured, so that several levels of difficulty adapt to the child's motor development. This app allows to easily manage users and record session results so that the therapist can perform a complete analysis considering different parameters such as accuracy and time.

This work is organized as follows: Section 2 presents materials and methods, Section 3 shows the main results obtained using this app, and Section 4 is devoted to conclusions.

2. Materials and Methods

The children sample was randomly selected among students from an education centre in the area of A Coruña involved in the program of detection and promotion of motor coordination for which the application was designed. Only those children who had not presented any problem at the motor level in the previous evaluations were included, with the aim of knowing the results of this software use on the general population of early school years (six primary school years, from 7 to 12 years, with 7–8 children per year). The app considers three difficulty levels associated with three movement paths (as shown in Figure 1) of increasing difficulty in terms of motor coordination i.e., straight line path, curved line path and zigzag line path. Each level is in turn subdivided into three modes according to the characteristics of these paths: slow speed (3 cm/s from first to third year and 4 cm/s from fourth to sixth year), fast speed (6 cm/s from first to third year and 8 cm/s from fourth to sixth year) and fast speed with disappearance of the moving object (with the same speeds as in the previous case).

The procedure for the experiment was the following. Firstly, we have to say that the evaluations on each subject were individually conducted by a physical therapist during Physical Education sessions. By moving an object along a path on the screen of the mobile device, as depicted in Figure 1, they could carry out the spatial prediction of movement activities to be then performed in those physical sessions. The tests were carried out in an empty classroom, with a relaxed sitting position and resting the mobile device, a tablet in this case, on a table. The estimated duration of the experiment per child was of 25 min, distributed as follows: the app was explained in the first 15 min, when was emphasized that the ball should be pressed on the screen to start each exercise and that the finger should not be separated from it at any time during the path to its end, all the levels and the different progressions of complexity within each one were explained, and the last minutes of this phase were devoted to practice the exercises of each level in a guided way, while in the final 10 min data was recorded until each child finished three complete repetitions with a permissible error doing all exercises contained in each of those 9 sub-levels of work.

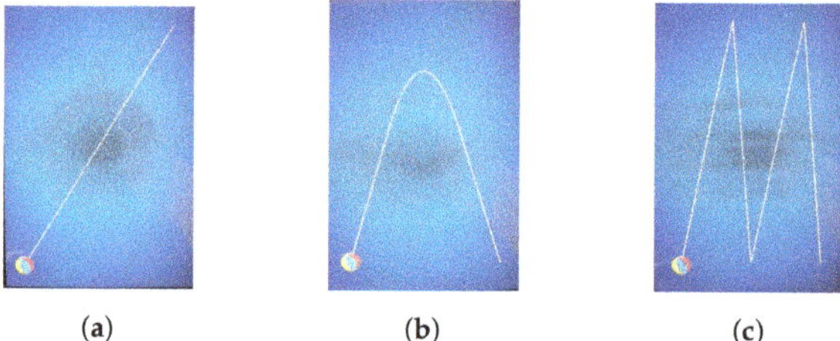

(a) (b) (c)

Figure 1. Types of movement paths: (**a**) Straight line path. (**b**) Curved line path. (**c**) Zigzag line path.

3. Results

From the data registered by our app, a first survey has been carried out considering motor coordination. An error measurement to quantify this skill is given by the euclidean distance between the path displayed on the mobile screen and what the user does, expressed in cm.

Figure 2a,c show box diagrams of the errors made by students of different primary school years for the three modes: slow speed, fast speed and fast speed without path display, respectively. The data corresponding to all different levels, i.e., the path types of Figure 1 and attempts made by each student, have been grouped together. For all figures, a clear error decrease can be observed as the school year increases. Between the third and the fourth year the differences are less noticeable, due to the higher speed used for the same mode in the last three years. Moreover, the error always increases with the speed (Figure 2a,b) and even more when the path display is hidden (Figure 2c). Figure 3a,c show box diagrams for levels 1, 2 and 3, respectively corresponding to the paths depicted in Figure 1, this time grouping the data corresponding to the different modes and attempts of each student. In this case you can see how the errors increase when considering increasingly complex movement paths. Again, it is observed that the error decreases as the course increases, except between the third and the fourth primary school year, for the same reason as explained above.

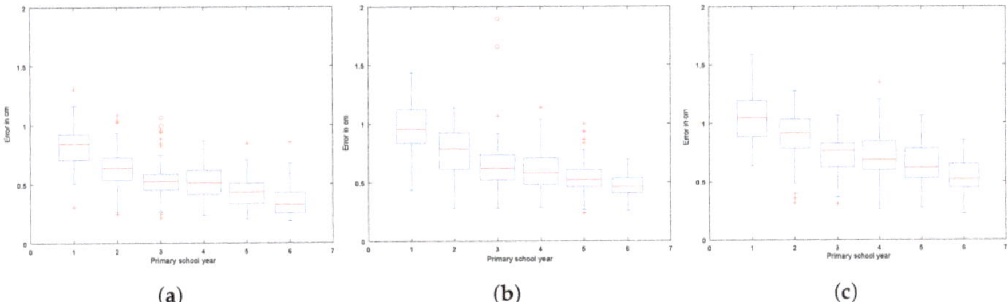

(a) (b) (c)

Figure 2. Motor coordination errors for three modes: (**a**) Slow speed. (**b**) Fast speed. (**c**) Fast speed without path display.

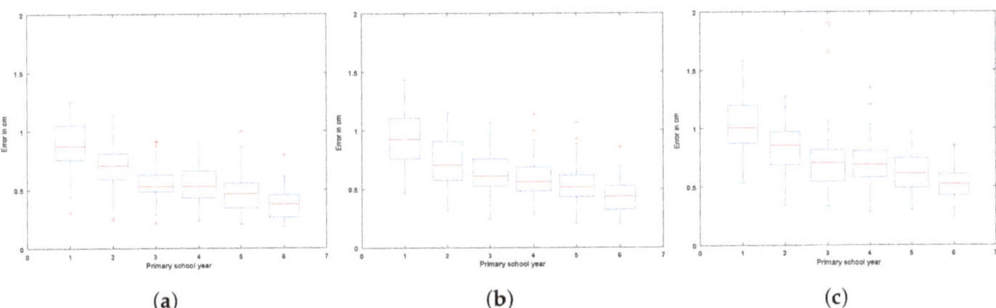

(a) (b) (c)

Figure 3. Motor coordination errors for three levels: (**a**) Straight line path. (**b**) Curved line path. (**c**) Zigzag line path.

4. Discussion and Conclusions

There are no similar tools designed for this purpose with which to compare these results. This application provides objective and measurable data on the accuracy and quality of the requested movement, which are ignored by current evaluation tools. In conclusion, it shows predictable behaviour in scenarios of healthy population, so that the following step is its use on children affected by motor coordination difficulties as support tool for their improvement and progress monitoring.

Author Contributions: F.J.V.-A. has implemented the app software; D.M.N. performed the experiments; D.M.N. and P.M.C. analyzed the data; J.V.C. and L.G.D. designed the experiments and P.M.C. and A.D. head the research.

Funding: This work has been funded by the Xunta de Galicia (ED431G2019/01), the Agencia Estatal de Investigación of Spain (TEC2016-75067-C4-1-R) and ERDF funds of the EU (AEI/FEDER, UE).

Conflicts of Interest: The authors declare no conflict of interest.

References

1. Gallahue, D.L.; Ozmun, J.C.; Goodway, J. *Understanding Motor Development: Infants, Children, Adolescents, Adults*; Mcgraw-Hill: Boston, MA, USA, 2006.
2. Skinner, R.A.; Piek, J.P. Psychosocial implications of poor motor coordination in children and adolescents. *Hum. Move. Sci.* **2001**, *20*, 73–94.
3. Jeng, Y.L.; Wu, T.T.; Huang, Y.M.; Tan, Q.; Yang, S.J. The add-on impact of mobile applications in learning strategies: A review study. *J. Educ. Technol. Soc.* **2010**, *13*, 3–11.

Proceedings

A-Frame as a Tool to Create Artistic Collective Installations in Virtual Reality †

João Azevedo [1,*], Paulo Veloso Gomes [2], João Donga [1] and António Marques [2]

[1] LabRP, Psychosocial Rehabilitation Laboratory, School of Media Arts and Design, Polytechnic Institute of Porto, 4200-072 Porto, Portugal; jpd@esmad.ipp.pt

[2] LabRP, Psychosocial Rehabilitation Laboratory, School of Allied Health Technologies, Polytechnic Institute of Porto, 4200-072 Porto, Portugal; pvg@ess.ipp.pt (P.V.G.); ajmarques@ess.ipp.pt (A.M.)

* Correspondence: joaoazevedo@esmad.ipp.pt

† Presented at the 3rd XoveTIC Conference, A Coruña, Spain, 8–9 October 2020.

Published: 26 August 2020

Abstract: Virtual Reality, due to its complexity and technological requirements, has a set of frictions that hinder its dissemination. The main ones can be summarized in the requirement to learn complex developing environments, like game engines, then we need to install applications, specific to each operating system and according to the means through which they can be accessed.

Keywords: virtual reality; A-Frame; WebVR; digital art; immersive environments

1. Introduction

The A-Frame framework, published in 2015 by Mozilla, aims to facilitate access to Virtual Reality (VR), namely through Internet browsers and all types of VR equipment. This objective is made possible by the structure on which it was developed, WebVR. It allows the creation of immersive environments that can be inhabited simultaneously by several users, even if access is made through different types of VR devices, from smartphones and personal computers to specific VR equipment, such as the Oculus Quest, Rift and HTC Vive.

As artists seek ways to express their creativity in VR, A-Frame can be a way to start creating immersive experiences without the need to learn a complex programming language or a game engine like Unity or Unreal Engine [1]. Plus, the A-Frame end results are easily delivered through the world wide web. This article focuses on the A-Frame potential to create, publish and share collective experiences of virtual immersive artistic installations. This work will demonstrate this potential through a teaching approach to this framework, and its results came from Bachelor of Arts (BA) and Master of Arts (MA) degree curricular units in a Media Art School.

2. Materials and Methods

The A-Frame core structure is a JavaScript API, an acronym for Application Programming Interface, that allows access to VR equipment through recent and popular Internet browsers, namely Mozilla Firefox, Google Chrome and Microsoft Edge. A-Frame provides a set of native elements that allow the creation of primitive geometric objects, lighting and audio systems. It is possible to import various types and formats of multimedia files, namely images, textures, videos and 3D models. All of these elements can be programmatically controlled by a simple markup language [2].

To write this code, the artist only needs a text editor; however, it is also recommended that an Integrated Development Environment (IDE) allowing the setup of a client side server is used in order to test the results locally and avoiding having to push it to an online server. However, if the developer needs to test the result online or share the code with other artists/developers, there are several platforms available. In the class we used Glitch and GitHub, the first was chosen for the first lessons due to its ease of use and code sharing features, the second was used in the advanced classes for the version control feature and the robust file uploader.

The fundamentals of A-Frame can be divided in four chapters: Geometry, Lightning, Animation and Interaction [3]. The classes followed this sequence since each one of these aspects grows in complexity.

In Geometry, we have the possibility to use predefined array primitives like a Box or Sphere. Each primitive has properties that define its dimension, position and materials.

The Lightning system controls the light and shadows of the scene and objects. These entities share some of generic properties of the primitives but then have the specific ones that define the type of light and shadow behavior.

The Animation component controls the movement of Geometry and Lightning entities over time. The Interaction in A-Frame is event-driven and its capable to handle keyboard/mouse, 3 and 6 degrees of freedom controllers.

3. Results

The artists of the MA degree had a creative approach but based on recognizable references, as shown in Figure 1.

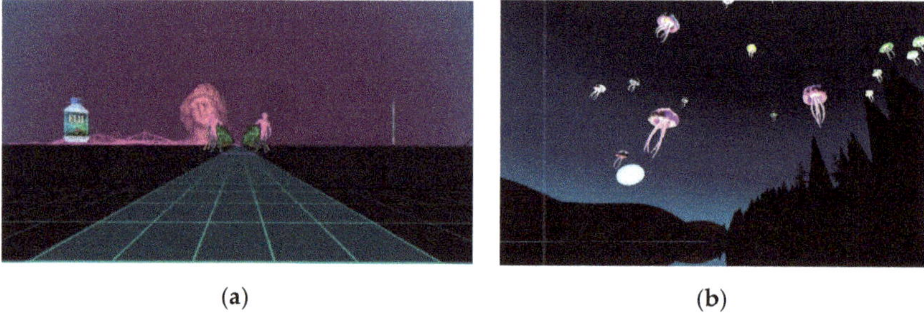

(**a**) (**b**)

Figure 1. (**a**) Vaporscape: A Jorge Barros project inspired by the art of the album "Floral Shoppe" by Vektroid; (**b**) A WebVR adaptation of the collective TeamLab "Forest of Resonating Lamps—One Stroke" by Luisa Guedes.

Despite the fact both courses had access to the same artistic references, most of the projects from the BA degree artists were related to the videogame culture (Figure 2). All artists from both courses fully explored the Geometry, Lighting and Animation properties, but only 8 out of 31 projects applied interaction to their creations.

(**a**) (**b**)

Figure 2. (**a**) Beacon is a project by Pedro Ermida and Rúben Ferreira; (**b**) A WebVR interpretation of the Super Mario environment by Luis Alves.

The results demonstrate that it is possible to create rich environments that can be shared over the Internet with a small learning curve when compared to game engines. Future work can evaluate the variance on the time needed to concept, develop and distribute this projects with other creative platforms.

Author Contributions: Conceptualization, J.A.; methodology, J.A. and P.V.G.; writing—review and editing, J.A.; P.V.G. and J.D.; supervision, A.M. and J.D.; project administration, A.M. and J.A. All authors have read and agreed to the published version of the manuscript.

Funding: This research received no external funding.

Conflicts of Interest: The authors declare no conflict of interest.

References

1. Ma, X.; Cackett, M.; Park, L.; Chien, E.; Naaman, M. Web-Based VR Experiments Powered by the Crowd. In Proceedings of the 2018 World Wide Web Conference, Lyon, France, 23–27 April 2018; pp. 33–43, doi:10.1145/3178876.3186034.
2. Santos, S.G.; Cardoso, J.C.S. A-Frame Experimentation and Evaluation for the Development of Interactive VR: A Virtual Tour of the Conimbriga Museum. 2019. Available online: http://hdl.handle.net/10316/87028 (accessed on 23 June 2020).
3. Aframe. A-Frame—Make WebVR. Available online: https://aframe.io/ (accessed on 23 June 2020).

Proceedings

Artificial Intelligence in Pre-University Education: What and How to Teach †

Sara Guerreiro-Santalla *, Francisco Bellas and Richard J. Duro

Grupo Integrado de Ingeniería, CITIC Research Center, Universidade da Coruña, 15403 Ferrol, Spain; francisco.bellas@udc.es (F.B.); richard@udc.es (R.J.D.)

* Correspondence: sara.guerreiro@udc.es

† Presented at the 3rd XoveTIC Conference, A Coruña, Spain, 8–9 October 2020.

Published: 26 August 2020

Abstract: The present paper is part of the European Erasmus+ project on educational innovation led by the UDC and entitled "AI+: Developing an Artificial Intelligence Curriculum adapted to European High School". In this paper, the progress achieved during the first year of the project will be presented. Mainly, the definition of the methodological approach for this future subject has been defined, and the AI topics to be dealt with at this age have been established. It has been a great effort to select the most appropriate focus for this subject considering the students' and teachers' technical background and the schools' equipment.

Keywords: artificial intelligence in education; STEM education; applied artificial intelligence; smartphone-based education

---

1. Introduction

The current society is starting to be connected to Artificial Intelligence (AI). Smartphones, social networks, voice assistants, are robots are just a few examples of AI devices used every day. As a consequence, the European Commission, as many other governments around the world, has initiated the creation of an AI plan for the EU states, in order to regulate how the transition to this new society will be performed [1]. In this process, education is a key field to assure that future generations are prepared for the new challenges.

Artificial intelligence has been a university subject for more than 30 years. Even so, it has never been included in official pre-university studies because it requires a technical background that was not acquired in secondary school, and the existing tools were not suitable for students at this level, until now. Currently, students have a good technological base thanks to subjects such as IT and programming. Moreover, AI systems have improved remarkably in the last decade, so now, students are not required to be technological experts in order to use them. An example of this is the TensorFlow Playground [2], which is an interactive neural network viewer that allows users unfamiliar with high-level coding to experiment with neural networks, considered a very advanced topic a few years ago.

Therefore, AI teaching in pre-university education is currently feasible, but the specific topics that must be taught at this age, and to what extent, is an open issue that must be faced. This is where the AI+ project comes in, as in the first year of the project's development, the methodological approach for this future subject has been defined, and the AI topics to be addressed have been established.

2. Teaching Organization and Methodology

The AI teaching approach to be followed here focuses on embedded intelligence, i.e., the programming of real-world devices that interact with real environments. Because access to embedded systems from schools is not easy, the use of the student's mobile phone (smartphone) as the central technological element for all educational material to be developed has been established as a key premise in the project. Currently, smartphones can be considered as general public devices with a low cost. Moreover, they have the technological level required for AI teaching in terms of sensors, actuators, computing power, and communications; and they will have this in the future because they are continuously updated.

Furthermore, the subject follows a STEM methodology (Science, Technology, Engineering, Mathematics), which focuses on the integrated learning of different technical and scientific concepts through an engineering approach. AI is a technological field that fits perfectly in the STEM methodology, since it requires knowledge from different disciplines to solve problems, such as physics, mathematics, programming, and design.

Regarding the units, each one will present a challenge or project, following the PBL (Project Based Learning) and cooperative methodology, which students will have to solve, organized into groups, in a creative and practical way. This approach is based on totally proactive learning and through real-world problem solving (learning by doing), in accordance with the eminently practical approach of the curriculum.

3. AI Topics for Pre-University Education

To define the AI topics to be studied at this academic level, the main elements that make up an AI system, from an engineering point of view, were taken as a starting point. As shown in the diagram of Figure 1, an AI system is made up of five sub-systems linked to the sensing and actuation stages.

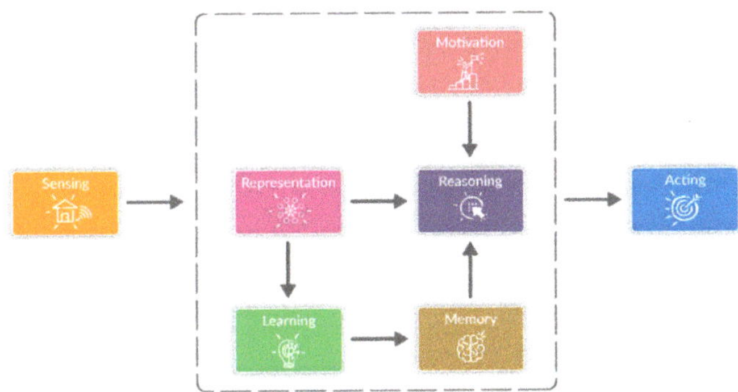

Figure 1. Main elements of an AI system.

Keeping this in mind, it has been decided that the eight topics to be addressed in the curriculum are: (1) perception, (2) actuation, (3) representation, (4) reasoning, (5) learning, (6) collective intelligence, (7) motivation, and (8) Sustainability, Ethics, and Legal aspects of AI (SEL). As can be seen, most of the implementation blocks displayed in Figure 1 correspond to specific topics that will be studied in the curriculum. It has been decided not to include the memory block as a topic and to add two new ones that are a fundamental part of AI from a didactic perspective: collective intelligence and SEL.

Firstly, it was decided that the first two topics would be those of perception and actuation, since they are part of all AI systems. This curriculum will focus on the sensors that are currently most used in AI, such as cameras, microphones, and touch screens. Regarding actuation, it will be focused on a more general approach beyond typical motors, and multiple actuators that can be found in AI systems, such as speakers or LCD screens, will be presented to students.

Representation and reasoning are two very relevant topics in AI, and completely new to students. For this level, basic concepts like topological or metric maps will be explained, and simple reasoning over graphs will be presented, included in the teaching units.

Learning is a key property of an AI system, which must be able to learn from its experience, generalizing the information it perceives. It is therefore a central issue to be studied in the curriculum, including supervised, unsupervised, and reinforcement learning fundamentals.

For the collective intelligence topic, students will learn how to communicate information between AI systems to improve their response and how to coordinate their actions to operate in a more reliable way.

Typically, in AI systems, the goal or objective to be achieved is imposed by the human designer. However, to obtain systems with a higher degree of autonomy, they must be able to discover their objectives and, in the case of having more than one, choose the best one at a given moment. That is what the motivation is focused on, and although it is an open issue in AI, students must understand how a motivated AI system will work and how it can be controlled by humans.

Finally, considering that AI's impact on different aspects of future societies will bring new problems, SEL is a topic that cannot be left aside in this curriculum.

4. Curriculum Organization

The AI curriculum will cover two academic courses. It has been decided to structure the curriculum to cover three fields of application in AI: intelligent smartphone apps, autonomous robotics, and smart environments. Although many other application fields could have been selected, these are very representative of the current AI domain, and all of them can be developed using a smartphone in classes.

To develop intelligent smartphone apps, it has been decided to use the App Inventor 2 environment [3]. In this interface, smart applications for the Android operating system can be developed using already existent AI modules, like [4]. Regarding autonomous robotics, the curriculum will include the use of the smartphone-based robot Robobo [5]. With it, students will practice their skills on most of the AI topics explained above, due to the technological capabilities of this platform [6]. Finally, for the smart environments field, it will be proposed to create an intelligent network with different devices, such as smartphones, smart speakers, and cameras, through which students will be able to create intelligent spaces and practice the collective intelligence topics in depth.

5. Conclusions

The topics to be included in a curriculum for teaching artificial intelligence in high schools and the most adequate methodology to present them to students have been defined as a part of the AI+ project. In the next two years, all the teaching units that conform to the curriculum will be developed and tested with students from the schools that are involved in the project.

Funding: This research was supported by the Erasmus+ program of the European Commission through Project Number 2019-1-ES01-KA201-065742.

Conflicts of Interest: The authors declare no conflict of interest.

References

1. Coordinated Plan on Artificial Intelligence (COM (2018) 795 Final). Available online: https://ec.europa.eu/knowledge4policy/publication/coordinated-plan-artificial-intelligence-com2018-795-final_en (accessed on 20 July 2020).
2. TensorFlow Playground. Available online: https://playground.tensorflow.org/ (accessed on 20 July 2020).
3. App inventor 2. Available online: https://appinventor.mit.edu/ (accessed on 22 July 2020).
4. Introduction to Machine Learning: Image Classification. Available online: https://appinventor.mit.edu/explore/ai-with-mit-app-inventor (accessed on 20 July 2020).

5. The Robobo Project. Available online: https://theroboboproject.com/ (accessed on 20 July 2020).
6. Bellas, F.; Naya, M.; Varela, G.; Llamas, L.; Prieto, A.; Becerra, J.C.; Bautista, M.; Faiña, A.; Duro, R.J. The Robobo Project: Bringing Educational Robotics Closer to Real-World Applications. In *Robotics in Education. RiE 2017. Advances in Intelligent Systems and Computing*; Springer: Cham, Switzerland; 2018; Volume 630.

Proceedings

Development of an Open Source Tool and a Multi-Platform for Generation of Forensic Reports †

Mireia Martí Vilas [1,*], Pilar Vila Avendaño [1,2] and José M. Vázquez-Naya [1,3]

1 Departamento de Computación, Facultad de Informática, Universidade da Coruña, Grupo RNASA-IMEDIR, Elviña, 15071 A Coruña, Spain; pilar.vila@forensic-security.com (P.V.A.); jose.manuel.vazquez.naya@udc.es (J.M.V.-N.)
2 Forensic & Security C/Copérnico, 3, 2ª planta, local A3, oficina 2, 15008 A Coruña, Spain
3 Centro de Investigación CITIC, Universidade da Coruña, Elviña, 15071 A Coruña, Spain
* Correspondence: mireia.marti@udc.es

† Presented at the 3rd XoveTIC Conference, A Coruña, Spain, 8–9 October 2020.

Published: 27 August 2020

Abstract: Computer Forensics is a science that is part of computer security and focuses on identifying, preserving, analyzing and presenting electronic evidence that has been found on a device. This process has to be thoroughly documented by the expert who carries it out, and must be adapted to standards such as UNE 197010:2015 or ISO/IEC 27042:2015. However, there are no tools to facilitate this task. Therefore, in this work, a multiplatform and open source tool is developed to facilitate the expert's elaboration of the report, and the management of the documentation related to the case, while keeping this information safe.

Keywords: computer forensics; forensic reports; cyber security; multi-platform; open source; forensics tool

1. Introduction

Computer forensics is a science that takes parts of informatics security and focuses on identifying, preserving, analyzing and showing electronic evidence that has been found in a device [1]. This process has to be carefully documented by an expert witness in order to describe the starting stage, check that evidence has not been manipulated, record the process followed and finally to write a conclusion. The documentation has to be in a report that, in this case, has to be carried to the court. For this reason, it is important that it is adapted to specific standards like rule UNE 197010:2015 ("General judgements for the production of reports and forensic expert opinions about Information Technologies and Communications") and the guide ISO/IEC 27042:2015 that lists specific instructions that must include the forensic report [1].

However, the legislation does not specify which tool or tools the expert witness has to use neither for the report redaction nor for the relative documentation management. Typically, the expert witness will make use of different tools to achieve this goal (word processor, file system of the OS itself to store the information, encryption tools to protect such information, etc.). However, this process is tedious, partly repetitive and prone to error. To automate this process, as far as possible, would be of great help to the expert witness, while providing greater certainty and more guarantees as to the correct wording of the report.

Due to the absence of specific tools for producing expert witness reports, this work is based on the development of a desktop, multi-platform, open-source application, which makes it easier for the expert witness to produce the report, while keeping the information safe.

2. Development

To achieve this development, an incremental development methodology has been followed, in which new functionalities have been added in each iteration. For the implementation we used the Python 3.0 language and the GTK+ library, allowing the result to be a multi-platform application.

The analysis and design deals with the corresponding use cases of each iteration, considering in each case the corresponding design decisions. Using the Balsamiq tool, the respective prototypes were created for each iteration.

The most basic functionalities of the application consist of adding and editing the expert cases. This process allows the user to fill in the fields corresponding to each phase of the expertise process (complying with the ISO/IEC 27042 guide), and also saving the case. For the storage, the user is offered the possibility of choosing the folder in which to store the case. Subsequently, a folder is created with the name of the case, in which all the documentation belonging to the case will be stored. The report will be stored in XML format to offer flexibility when exporting it to other formats.

A highlight of the storage process is the encryption of the information. Hybrid cryptography is used to store the encrypted report. This consists of using symmetric cryptography to encrypt the document (using a random key for each) and asymmetric cryptography to encrypt the random key (see Figure 1).

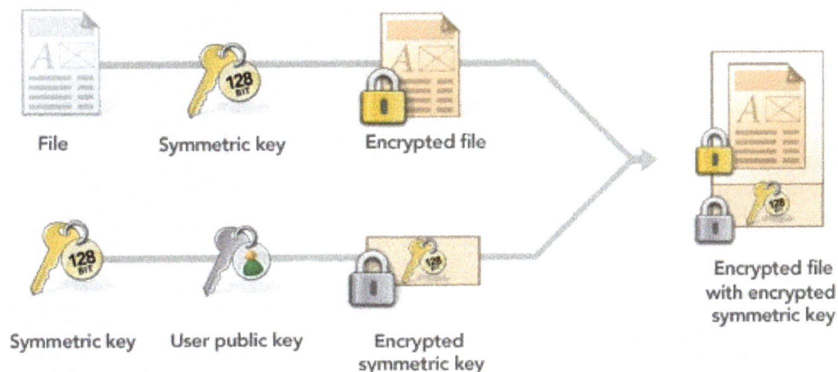

Figure 1. Hybrid cryptography scheme.

Another functionality of the tool is evidence management. To store the evidence in a folder within the case, the user can drag it into the application. These will be stored encrypted, and the name will be displayed for easy reference.

The main data of the expert should always go at the beginning of the report. Since these will not vary from one case to another, the tool allows the insertion (and editing) of the data independently of the case. In this way, the user will only have to introduce them once, and it will be the tool itself that will attach them to the beginning of each generated report.

Finally, the application implements authentication to be able to access the reports already generated. During the first access a new password is requested and the key pair (public and private) is generated. This password will be valid to store the private key in a safe way. This way, the password will be necessary to decrypt a report and be able to edit it. If for any reason the user forgets the password, it is possible to revoke it and create a new one (with its respective key pair), leaving the cases generated so far inaccessible.

3. Results and Conclusions

The drafting of the expert's report is one of the most important phases of the expertise. This tool offers functionalities such as password access, a clear and simple interface, the secure storage of the report through encryption and exporting the report to different formats, among others. In short, it offers the user an easy way to manage the report and all the documentation pertaining to an expert case, complying with the aforementioned standards.

Author Contributions: Conceptualization, M.M.V.; Methodology, M.M.V., P.V.A. and J.M.V.-N.; Software, M.M.V.; Investigation, M.M.V., P.V.A. and J.M.V.-N.; Resources, P.V.A. and J.M.V.-N.; Writing–original draft preparation, M.M.V.; Writing–review and editing, M.M.V., P.V.A. and J.M.V.-N.; Supervision, P.V.A. and J.M.V.-N. All authors have read and agreed to the published version of the manuscript.

Funding: This work was supported by the "Collaborative Project in Genomic Data Integration (CICLOGEN)" PI17/01826 funded by the Carlos III Health Institute from the Spanish National plan for Scientific and Technical Research and Innovation 2013-2016 and the European Regional Development Funds (FEDER)—"A way to build Europe". This project was also supported by the General Directorate of Culture, Education and University Management of Xunta de Galicia ED431D 2017/16 and "Drug Discovery Galician Network" Ref. ED431G/01 and the "Galician Network for Colorectal Cancer Research" (Ref. ED431D 2017/23), and finally by the Spanish Ministry of Economy and Competitiveness for its support through the funding of the unique installation BIOCAI (UNLC08-1E-002, UNLC13-13-3503) and the European Regional Development Funds (FEDER) by the European Union. Additional support was offered by the Consolidation and Structuring of Competitive Research Units—Competitive Reference Groups (ED431C 2018/49), funded by the Ministry of Education, University and Vocational Training of the Xunta de Galicia endowed with EU FEDER funds.

Conflicts of Interest: The authors declare no conflict of interest.

References

1. Avendaño, P.V. *Técnicas de Análisis Forense informático para Peritos Judiciales profesionales*; 0xWord: Madrid, Spain, 2018; pp. 15–27.

Proceedings

Identification of Prevotella, Anaerotruncus and Eubacterium Genera by Machine Learning Analysis of Metagenomic Profiles for Stratification of Patients Affected by Type I Diabetes †

Diego Fernández-Edreira [1], Jose Liñares-Blanco [1,2,*] and Carlos Fernandez-Lozano [1,2]

1 Department of Computer Science and Information Technologies, Faculty of Computer Science, Universidade da Coruña, Campus Elviña s/n, 15071 A Coruña, Spain; diego.fedreira@udc.es (D.F.-E.); carlos.fernandez@udc.es (C.F.-L.)

2 CITIC-Research Center of Information and Communication Technologies, Universidade da Coruña, 15071 A Coruña, Spain

* Correspondence: j.linares@udc.es; Tel.: +34-881-01-1302

† Presented at the 3rd XoveTIC Conference, A Coruña, Spain, 8–9 October 2020.

Published: 27 August 2020

Abstract: Previous works have reported different bacterial strains and genera as the cause of different clinical pathological conditions. In our approach, using the fecal metagenomic profiles of newborns, a machine learning-based model was generated capable of discerning between patients affected by type I diabetes and controls. Furthermore, a random forest algorithm achieved a 0.915 in AUROC. The automation of processes and support to clinical decision making under metagenomic variables of interest may result in lower experimental costs in the diagnosis of complex diseases of high prevalence worldwide.

Keywords: diabetes; machine learning; microbiome; metagenomics; data science

---

1. Introduction

It is known that diabetes type I (DTI) is a disease that is closely linked to changes in the microbiota [1]. Typically, works that study the metagenomic profile of a microbe in DTI uses only conventional statistical approaches [2]. Therefore, in this work a novel methodology to analyze DTI status using machine learning (ML) is proposed. In addition, new metagenomics genera are been identified with potential in the development of this disease.

2. Materials and Methods

OTUs genera faecal samples from 124 newborns were downloaded from Diabinmune project [2]. The experimental design starts removing near zero features and scaling the data; Random Forest (RF) [3] and glmnet [4] algorithms were used following a nested cross validation (CV) approach for training the models. A holdout was used for hyperparameter tuning (2/3 for training and 1/3 for testing) followed by a 10-fold CV for model validation (repeated 5 times).

3. Results

We have obtained 45 genera suitable for carrying out the study. Figure 1a showed the experimental results carried out. We found a statistical difference between the models and the best results were achieved with RF. Feature importance is shown in Figure 1b. *Prevotella* is the bacteria with the higher accumulated importance along with *Anaerotruncus*, *Scherichia*, *Eubacterium*, *Odoribacter* and *Collinsella*.

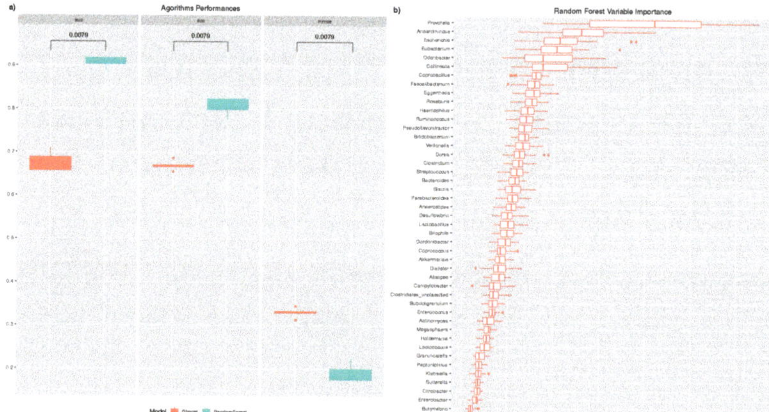

Figure 1. (**a**) Comparison of the 5 times 10-fold CV using a Wilcoxon test and (**b**) RF variable importance.

4. Discussion

We found in the literature that *Prevotella* and *Eubacterium* are strongly linked to DTI and *Anaerotruncus* with gestational diabetes. All of them are also correlated with instestinal dysbiosis processes [5,6]. In summary, we demonstrated the feasibility of a ML analysis of metagenomic profiles.

Author Contributions: Conceptualization, J.L.-B. and C.F.-L.; methodology, C.F.-L.; software, D.F.-E., J.L.-B. and C.F.-L.; formal analysis, D.F.-E.; Writing—Original Draft preparation, D.F.-E.; Writing—Review and Editing, D.F.-E., J.L.-B. and C.F.-L.; supervision, J.L.-B. and C.F.-L. All authors have read and agreed to the published version of the manuscript.

Funding: This work was supported by the "Collaborative Project in Genomic Data Integration (CICLOGEN)" PI17/01826 funded by the Carlos III Health Institute from the Spanish National plan for Scientific and Technical Research and Innovation 2013–2016 and the European Regional Development Funds (FEDER)—"A way to build Europe." and the General Directorate of Culture, Education and University Management of Xunta de Galicia (Ref. ED431G/01, ED431D 2017/16), the "Galician Network for Colorectal Cancer Research" (Ref. ED431D 2017/23) and Competitive Reference Groups (Ref. ED431C 2018/49). The funding body did not have a role in the experimental design; data collection, analysis and interpretation; and writing of this manuscript.

References

1. Tai, N.; Wong, F.S.; Wen, L. The role of gut microbiota in the development of type 1, type 2 diabetes mellitus and obesity. *Rev. Endocr. Metab. Disord.* **2015**, *16*, 55–65.
2. Kostic, A.D.; Gevers, D.; Siljander, H.; Vatanen, T.; Hyötyläinen, T.; Hämäläinen, A.M.; Peet, A.; Tillmann, V.; Pöhö, P.; Mattila, I.; et al. The Dynamics of the Human Infant Gut Microbiome in Development and in Progression toward Type 1 Diabetes. *Cell Host Microbe* **2016**, *20*, 121.
3. Breiman, L. Random forests. *Mach. Learn.* **2001**, *45*, 5–32.
4. Friedman, J.; Hastie, T.; Tibshirani, R. Regularization Paths for Generalized Linear Models via Coordinate Descent. *J. Stat. Softw.* **2010**, *33*, 1–22.
5. Siljander, H.; Honkanen, J.; Knip, M. Microbiome and type 1 diabetes. *EBioMedicine* **2019**, *46*, 512–521.
6. Hasan, S.; Aho, V.; Pereira, P.; Paulin, L.; Koivusalo, S.B.; Auvinen, P.; Eriksson, J.G. Gut microbiome in gestational diabetes: a cross-sectional study of mothers and offspring 5 years postpartum. *Acta Obstet. Gynecol. Scand.* **2018**, *97*, 38–46.

Proceedings

aspBEEF: Explaining Predictions Through Optimal Clustering †

Pedro Cabalar [1], Rodrigo Martín [1], Brais Muñiz [1,2,*] and Gilberto Pérez [1,2]

1 Department of Computer Science, University of A Coruña, 15001 A Coruña, Spain; pedro.cabalar@udc.es (P.C.); r.martin1@udc.es (R.M.); gilberto.pvega@udc.es (G.P.)
2 CITIC Research Center, University of A Coruña, 15001 A Coruña, Spain
* Correspondence: brais.mcastro@udc.es
† Presented at the 3rd XoveTIC Conference, A Coruña, Spain, 8–9 October 2020.

Published: 28 August 2020

Abstract: In this paper we introduce aspBEEF, a tool for generating explanations for the outcome of an arbitrary machine learning classifier. This is done using Grover's et al. framework known as Balanced English Explanations of Forecasts (BEEF) that generates explanations in terms of in terms of finite intervals over the values of the input features. Since the problem of obtaining an optimal BEEF explanation has been proved to be NP-complete, BEEF existing implementation computes an approximation. In this work we use instead an encoding into the Answer Set Programming paradigm, specialized in solving NP problems, to guarantee that the computed solutions are optimal.

Keywords: knowledge representation; answer set programming; explainable AI

---

1. Introduction

Explainability is one of the clear barriers Machine Learning (ML) has not yet overcome. In some domains, such as medicine, where decisions can seriously affect people's lives, experts need to understand the decisions of ML models, to avoid biased or unfair decisions. Unfortunately, explaining why a classifier makes a given prediction is a non-trivial task, specially in human terms. Explainable Artificial Intelligence (XAI) field studies tools that aim to explain the predictions of ML models by externally analyzing them, treating models as black boxes. This approaches are independent of the algorithm's design and would be able to explain already deployed models retrospectively. One example of such a tool is the framework *Balanced English Explanations of Forecasts* (BEEF) [1] which produces natural language explanations of an ML model by grouping theirs predictions into clusters. However, the BEEF process of finding such clusters is heuristic-guided and does not provide the optimal solutions. In this paper we present a prototype called aspBEEF, which grants the set optimal of BEEF clusters through the use of Answer Set Programming (ASP) [2–4]. In the following text we first explain in more detail how BEEF computes its explanations. Then, we present how our prototype aspBEEF works. After that, we show a short evaluation of the prototype. Finally, we comment about the future work and conclude the paper.

2. BEEF Computation of Clusters

BEEF is a framework to explain the outcome of any ML binary classifier in terms of intervals over the input features. Given a set of predictions, BEEF first clusters them using some traditional method (such as KMeans). The result of the clustering is used as a starting point to find a set of *axis-aligned, hyperrectangular clusters* (boxes for short) that satisfy some previously given thresholds in terms of (1) purity: the majority predicted class in each cluster; (2) overlapping: the amount of space shared by several clusters; and (3) inclusion: the lack of predictions outside any cluster. The algorithm iteratively adjusts the boundaries of the boxes, until the thresholds are satisfied. This problem has been proved to be NP-complete by Grover et al. [1]. Their framework uses some heuristics during the search, at the cost of losing optimality.

Boxes are *axis-aligned* and finite, so they can be described as a set of intervals over each dimension (each input feature). Each box is labeled with the name of the majority predicted class within. We can explain a model outcome o just finding out the description (the intervals) of the boxes o fell in. The descriptions of those boxes whose label matches the predicted class will be the *supporting explanations*. The rest will be the *opposing explanations*. The sum of supporting and opposing explanations is what BEEF calls a balanced explanation.

3. The `aspBEEF` Tool

For finding the optimal solutions, `aspBEEF` makes use of ASP, a declarative Knowledge Representation and problem-solving paradigm. Under ASP, the domain knowledge and the problem to solve must be specified as a set of rules in a logic program, from which a solver obtain the solutions in terms of models of the program called answer sets. ASP also allows optimization among answer sets in different ways: in this work, we use the ASP extension `asprin` [5] that allows a flexible way of defining preference relations for optimization. In particular, the use of `asprin` preference relations enables us to shift between simple and general explanations or fitted and complex ones depending on the problem.

`aspBEEF` takes the ML model predictions, the number of clusters and the feature configuration to generate the BEEF balanced explanations. The tool provides those explanations as a set of rules in ASP or as a graphical representation. Since development is not closed yet, there exist some differences with BEEF. While BEEF maximizes inclusion, `aspBEEF` minimizes exclusion since ASP handles minimization easier. Furthermore, BEEF clusters have their own set of active dimensions while in `aspBEEF` the dimensions are active or inactive globally. Finally, `aspBEEF` allows the user to choose the number of different features to use, some of those can be fixed to reduce the search space.

4. Evaluation

Evaluation has been done by running `aspBEEF` with random samples of three different sizes of the publicly available IRIS data set (https://archive.ics.uci.edu/ml/datasets/iris) and using both the fixed and free feature selection methods to compare performance. Each measure has been taken 100 times and then averaged to smooth out outlying values, as ASP solving is a non-deterministic algorithm.

The search space and the computational time grows exponentially with the number of free features. Cases in which all of the features are selected, (e.g., all of the four features in the case of Table 1) eliminate any decision-making over feature selection completely, thus greatly reducing the problem complexity. The best times are achieved using a pre-fixed set of features, but this requires previous knowledge of the search space.

Despite the long computational times when using free features, automatic feature selection provides additional insight about the explanations given by the system, as it prunes the less useful features. Nevertheless, performance is acceptable even for the most complex case-scenarios.

Table 1. Times table for a data set of 150 points and 4 features.

Sample Size	Used Features	Time w/ Free Features	Time w/ Fixed Features
60	2	1.1288	0.7253
60	3	1.1103	0.6995
60	4	0.5443	0.5700
90	2	3.4666	1.6238
90	3	2.5741	1.3963
90	4	1.3596	1.1760
150	2	27.7299	5.5057
150	3	28.8644	5.9140
150	4	7.7072	6.1569

5. Conclusions and Future Work

We have introduced the tool aspBEEF whose main feature is to obtain optimal explanations for the BEEF framework. Evaluation shows that the technique has a high computational cost. To solve this problem, feature selection should be performed externally, to fix the important features. Anyway the effort is reasonable having in mind that our goal is not simple classification but obtaining rich explanations of the outcome of any ML classifier.

As future work, we aim to implement BEEF feature selection, modifying the activation of features to be cluster-dependent. Furthermore, we want to improve flexibility by adding an option to make aspBEEF find the best number or rectangles, instead of being a user's choice.

The aspBEEF prototype is open source and already available at https://github.com/Trigork/aspBEEF.

Funding: This work has been partially supported by MINECO (grant TIN2017-84453-P) and Xunta de Galicia (grants GPC ED431B 2019/03 and ED431G 2019/01 for CITIC center).

References

1. Grover, S.; Pulice, C.; Simari, G.I.; Subrahmanian, V.S. BEEF: Balanced English Explanations of Forecasts. *IEEE Trans. Comput. Soc. Syst.* **2019**, *6*, 350–364. doi:10.1109/TCSS.2019.2902490.
2. Niemelä, I. Logic Programs with Stable Model Semantics as a Constraint Programming Paradigm. *Ann. Math. Artif. Intell.* **1999**, *25*, 241–273.
3. Marek, V.W.; Truszczyński, M. Stable Models and an Alternative Logic Programming Paradigm. In *The Logic Programming Paradigm*; Apt, K.R., Marek, V.W., Truszczyński, M., Warren, D., Eds.; Artificial Intelligence, Springer: Berlin/Heidelberg, Germany, 1999; pp. 375–398.
4. Brewka, G.; Eiter, T.; Truszczynski, M. Answer set programming at a glance. *Commun. ACM* **2011**, *54*, 92–103. doi:10.1145/2043174.2043195.
5. Brewka, G.; Delgrande, J.P.; Romero, J.; Schaub, T. asprin: Customizing Answer Set Preferences without a Headache. In Proceedings of the Twenty-Ninth AAAI Conference on Artificial Intelligence, Austin, TX, USA, 25–30 January 2015; pp. 1467–1474.

Proceedings

Mining of the Milky Way Star Archive Gaia-DR2. Searching for Binary Stars in Planetary Nebulae †

Iker González-Santamaría [1,2,*], Minia Manteiga [2,3], Carlos Dafonte [1,2], Arturo Manchado [4,5] and Ana Ulla [6]

1 Department of Computer Science and Information Technology, Universidade da Coruña (UDC), Campus Elviña sn, 15071 A Coruña, Spain; carlos.dafonte@udc.es

2 CITIC, Centre for Information and Communications Technology Research, Universidade da Coruña (UDC), Campus de Elviña sn, 15071 A Coruña, Spain; manteiga@udc.es

3 Department of Nautical Sciences and Marine Engineering, Universidade da Coruña (UDC), Paseo de Ronda 51, 15011 A Coruña, Spain

4 Instituto de Astrofísica de Canarias, 38200 La Laguna, Tenerife, Spain; amt@iac.es

5 Astrophysics Department, Universidad de La Laguna (ULL), CSIC, 38206 La Laguna, Tenerife, Spain

6 Applied Physics Department, Universidade de Vigo (UVIGO), Campus Lagoas-Marcosende, s/n, 36310 Vigo, Spain; ulla@uvigo.es

* Correspondence: iker.gonzalez@udc.es

† Presented at the 3rd XoveTIC Conference, A Coruña, Spain, 8–9 October 2020.

Published: 31 August 2020

Abstract: The aim of this work is to search for binary stars associated to planetary nebulae (ionized stellar envelopes in expansion), by mining the astronomical archive of Gaia DR2, that is composed by around 1.7 billion stellar sources. For this task, we selected those objects with coincident astrometric parameters (parallaxes and proper motions) with the corresponding central star, among a sample of 211 planetary nebulae. By this method, we found eight binary systems, and we obtained their components positions, separations, temperatures and luminosities, as well as some of their masses and ages. In addition, we estimated the probability for each companion star of having been detected by chance and we analyzed how the number of false matches increase as the separation distance between both stars gets larger. All these procedures have been carried out making use of data mining techniques.

Keywords: Gaia DR2; planetary nebulae; binary stars; astrometry

---

1. Introduction

This work is focused on the study and analysis of planetary nebulae (PNe), the stellar objects that are generated when stars of low and intermediate mass, reaching their final phase of evolution, ionize the envelope that surrounds them. For this task, we have used the Gaia Data Release 2 (GDR2) database, which contains astrometric and photometric parameters of around 1.7 billion stellar objects, so its mining requires the use of big-data techniques. Gaia is an ESA satellite that was put into orbit in late 2013 with the aim of making a star map of the Milky Way.

The initial objective of our study was to generate a catalog of PNe, starting from different bibliographic sources and identifying their corresponding central stars (CSs) in Gaia DR2. To carry out this data mining process, we performed a cross-match between the known coordinates of the PNe and the coordinates of the sources in the Gaia DR2 database. As a result, a total of 1571 CSs with known parallaxes (angle of variation in the sky when observing an object between two opposite times of the year) were identified. From the parallaxes we can derive the distances at which the objects are. To do this, we use a Bayesian statistical approximation method, which assumes an exponentially decreasing probability density space for distances. Knowing the distance to the PNe has a great importance, since it allows us to calculate the intrinsic parameters of its CS from the observational ones. For a more detailed study, we selected those objects with the best astrometric accuracy, we called this sample the Golden Astrometry Planetary Nebulae (GAPN).

The Gaia archive also allows to search for stars gravitationally linked to the CSs of these PNe, which would form a binary star system. The presence of binary stars in PNe allow us to better understand the formation and evolution of this late stellar evolutionary phase, Boffin [1]. It can also shed some light on its relationship with the aspherical morphologies of PNe.

2. Materials and Methods

The extremely precise measurements of parallaxes and proper motions (velocities in the plane of the sky) in Gaia DR2 has allowed us to search for co-moving stars in a region around each CS of the GAPN sample. Thus, we analyzed all the objects around each CS, and we selected those with coincidence in the astrometric parameters (parallaxes and proper motions) within their measurement errors, which would allow them to be classified as co-moving stars. To discard random coincidences, the probability density of the astrometric values in each field around the CSs was calculated, and the probability of finding an object with such coincident values with those of the CS was calculated, obtaining very low, practically negligible probabilities.

In addition, we also did an analysis about how the number of false matches increases as a function of the increase in the separation distance between both stars. To estimate this number of false matches, we analyzed an adjacent region to that of the CS with the same searching radius as the original region, and we obtained the number of co-moving objects to the CS in this area. By plotting the distribution of these false matches against the separation distance between the two stars, we were able to determine a maximum separation value for our selection of binary systems.

3. Results

By applying this searching procedure, we have been able to find eight binary systems as CSs of the PNe in our sample, with projected separations of less than 15,000 AU. One of them could even be a triple system. We have determined some astrometric parameters of these binary systems, as positions, proper motions or angular and physical separations. All these values can be consulted in Table 1.

Table 1. Astrometric data of the central and companion stars of each system.

Object	RA (º)	Dec (º)	Parallax (mas)	Distance (pc)	Sep. (AU)	PM_{RA} (mas/yr)	PM_{Dec} (mas/yr)
Abell 24 (CS)	117.9065	3.0059	1.40 ± 0.15	691^{+70}_{-59}	-	-4.37 ± 0.23	-0.75 ± 0.14
Abell 24 (B)	117.9067	3.0021	1.32 ± 0.07	725^{+29}_{-27}	9912	-4.25 ± 0.11	-0.97 ± 0.09
Abell 33 (CS)	144.788	−2.8084	1.01 ± 0.10	932^{+77}_{-67}	-	-14.76 ± 0.17	9.42 ± 0.15
Abell 33 (B)	144.7878	−2.8089	1.11 ± 0.09	856^{+57}_{-51}	1542	-14.94 ± 0.15	9.62 ± 0.14
Abell 34 (CS)	146.3973	−13.1711	0.83 ± 0.12	1118^{+144}_{-115}	-	3.08 ± 0.19	-9.13 ± 0.24
Abell 34 (B)	146.3954	−13.1693	0.81 ± 0.06	1155^{+56}_{-51}	10,472	3.23 ± 0.09	-9.11 ± 0.11
NGC 246 (CS)	11.7639	−11.872	1.92 ± 0.11	506^{+70}_{-59}	-	-16.96 ± 0.22	-8.88 ± 0.13
NGC 246 (B)	11.7647	−11.8727	1.77 ± 0.06	547^{+10}_{-10}	2118	-16.61 ± 0.09	-8.76 ± 0.08
NGC 3699 (CS)	171.991	−59.9579	0.62 ± 0.11	1506^{+279}_{-205}	—	-3.19 ± 0.16	1.14 ± 0.15
NGC 3699 (B)	171.9922	−59.9585	0.58 ± 0.07	1571^{+146}_{-123}	5036	-3.22 ± 0.10	1.07 ± 0.10
NGC 6853 (CS)	299.9016	22.7212	2.63 ± 0.06	372^{+6}_{-6}	-	10.39 ± 0.09	3.66 ± 0.09
NGC 6853 (B)	299.9005	22.7197	2.56 ± 0.06	382^{+7}_{-6}	2453	10.22 ± 0.09	3.81 ± 0.09
NGC 6853 (C)	299.8997	22.7202	2.30 ± 0.52	457^{+181}_{-101}	3322	9.13 ± 0.64	3.78 ± 0.78
PHR J1129-6012 (CS)	172.4594	−60.2022	0.41 ± 0.09	2159^{+448}_{-320}	-	-6.64 ± 0.14	2.34 ± 0.12
PHR J1129-6012 (B)	172.4573	−60.2022	0.35 ± 0.09	2482^{+576}_{-400}	9551	-6.84 ± 0.38	2.70 ± 0.12
PN SB 36 (CS)	268.5868	−39.1772	0.57 ± 0.08	1610^{+192}_{-156}	-	4.43 ± 0.11	-4.75 ± 0.10
PN SB 36 (B)	268.5831	−39.1761	0.70 ± 0.08	1331^{+128}_{-108}	14,880	4.43 ± 0.10	-4.78 ± 0.09

In addition, by obtaining the photometry of the companion stars, we were able to adjust their spectral energy distributions (SED) to theoretical models and then estimate their temperatures and luminosities. Moreover, by locating the companion stars in a color-magnitude diagram together with the PARSEC isochrones in Gaia DR2 passbands (Evans [2]), we could estimate some of their evolutionary ages and masses. Thus, we had the possibility to compare all these parameters with the CSs ones, obtained in our previous work, González-Santamaría [3].

4. Discussion

By using data mining methods and the accurate astrometry of Gaia, we have been able to select a small sample of CSPNe, with known and precise parallaxes, among a total of around 1.7 billion sources in Gaia DR2 database. Then, we have had the possibility to estimate their distances, and this knowledge has enabled us to derive their intrinsic properties from the observational ones.

Furthermore, the combination of the parallaxes with the proper motions has allowed us to search for companion stars to the CSs. We have applied quite strict selection criteria in order to avoid selecting any false co-moving object. We also have used a big-data method to estimate the probability of having select by chance any of the companion stars, by analysing the probability distribution of the astrometric parameters in the field around each of the CSs.

In the near future, with the launching of Gaia DR3, it is expected to be more quantity of astrometric data and with more accuracy. This will allow us to detect more binary systems associated to PNe.

Author Contributions: Conceptualization, M.M. and A.M.; methodology, M.M., A.M. and I.G.-S.; software, I.G.-S.; investigation, I.G.-S., M.M. and A.M; writing–original draft preparation, I.G.-S. and M.M.; writing–review and editing, I.G.-S., M.M, C.D., A.M. and A.U.; supervision, M.M. and C.D. All authors have read and agreed to the published version of the manuscript.

Funding: Funding from Spanish Ministry projects ESP2016-80079-C2-2-R, RTI2018-095076-BC22, Xunta de Galicia ED431B 2018/42, and AYA-2017-88254-P is acknowledged by the authors. M.M. thanks the Instituto de Astrofísica de Canarias for a visiting stay funded by the Severo Ochoa Excellence programme. IGS acknowledges financial support from the Spanish National Programme for the Promotion of Talent and its Employability grant BES-2017-083126 cofunded by the European Social Fund.

Acknowledgments: This work has made use of data from the European Space Agency (ESA) Gaia mission and processed by the Gaia Data Processing and Analysis Consortium (DPAC). This research has made use of the SIMBAD database, HASH database, and the ALADIN applet.

Conflicts of Interest: The authors declare no conflict of interest.

References

1. Boffin, H.M.J.; Jones, D. *The Importance of Binaries in the Formation and Evolution of Planetary Nebulae*; Springer: Berlin/Heidelberg, Germany, 2019; doi:10.1007/978-3-030-25059-1.
2. Evans, D.W.; Riello, M.; De Angeli, F. Gaia Data Release 2. Photometric content and validation. *Astron. Astrophys.* **2019**, *616*, A4, doi:10.1051/0004-6361/201832756.
3. González-Santamaría, I.; Manteiga, M.; Manchado, A.; Ulla, A.; Dafonte, C. Properties of central stars of planetary nebulae with distances in Gaia DR2. *Astron. Astrophys.* **2019**, *630*, A150, doi:10.1051/0004-6361/201936162.

Proceedings

Study of Machine Learning Techniques for EEG Eye State Detection †

Francisco Laport *, Paula M. Castro, Adriana Dapena, Francisco J. Vazquez-Araujo and Daniel Iglesia

CITIC Research Center & University of A Coruña, Campus de Elviña, A Coruña 15071; paula.castro@udc.es (P.M.C.); adriana.dapena@udc.es (A.D.); francisco.vazquez@udc.es (F.J.V.-A.); daniel.iglesia@udc.es (D.I.)

* Correspondence: francisco.laport@udc.es

† Presented at the 3rd XoveTIC Conference, A Coruña, Spain, 8–9 October 2020.

Published: 31 August 2020

Abstract: A comparison of different machine learning techniques for eye state identification through Electroencephalography (EEG) signals is presented in this paper. (1) Background: We extend our previous work by studying several techniques for the extraction of the features corresponding to the mental states of open and closed eyes and their subsequent classification; (2) Methods: A prototype developed by the authors is used to capture the brain signals. We consider the Discrete Fourier Transform (DFT) and the Discrete Wavelet Transform (DWT) for feature extraction; Linear Discriminant Analysis (LDA) and Support Vector Machine (SVM) for state classification; and Independent Component Analysis (ICA) for preprocessing the data; (3) Results: The results obtained from some subjects show the good performance of the proposed methods; and (4) Conclusion: The combination of several techniques allows us to obtain a high accuracy of eye identification.

Keywords: Discrete Fourier Transform; Discrete Wavelet Transform; Linear Discriminant Analysis; Support Vector Machine; Independent Component Analysis

1. Introduction

Over the past decades, several studies have proved that each eye state represents specific activity patterns in certain brain rhythms. For instance, it has been demonstrated that the alpha power increases during closed-eyes states while significant reductions are produced when subjects open their eyes. On the other hand, beta band power does not show relevant differences between both eye states [1]. Taking these studies into account, in our previous work [2] we proposed a threshold-based classification system for the detection of open eyes (oE) and closed eyes (cE) from Electroencephalography (EEG) signals. For this purpose, we employed the mean power ratio between alpha (8–12 Hz) and beta (13–17 Hz) frequency bands for the identification of the user's eye state. The system achieved an accuracy higher than 95 % for both eye states with a classification delay of 2 s.

In this paper, we extend our previous study by considering an EEG device with two sensors and different techniques for feature extraction, such as Independent Component Analysis (ICA), Discrete Wavelet Transform (DWT) and Discrete Fourier Transform (DFT). Moreover, two well-known classifiers for EEG such as Support Vector Machine (SVM) and Linear Discriminant Analysis (LDA) are also tested and assessed [3].

2. Material and Methods

The same EEG device described in [2] is employed to record the brain signals of the subjects using two electrodes, located at positions O1 and O2 according to the 10–20 International System. The experimental group included a total of 7 volunteers (males) who agreed to participate in the research. Their mean age was 29.67 (range 24–56). All of them indicated that they do not present hearing or visual impairments. Participation was voluntary and informed consent was obtained for each participant in order to employ their EEG data in our study.

A 10 min-recording was performed for each subject, where 5 min correspond to oE and the remaining 5 min to cE: 4 min are employed for the training step, where 2 of them correspond to the oE state and 2 to the cE one. Thus, the remaining 6 min of the recordings, composed by 3 min of each eye state, are employed for the test step.

The mean power of the alpha and beta bands, denoted by α and β, respectively, is computed for each sensor. According to these values, we extract their relative power ratio, given by $R = \beta/\alpha$, as a feature to predict the eye state. For this purpose, three different approaches are compared. On the one hand, the DFT is applied over the raw EEG data to compute α and β and their subsequent ratio. Moreover, as a second approach, the JADE algorithm [4] is used to extract the independent components from the original data previous to the feature extraction with the DFT. Finally, the DWT with 4 levels of decomposition and coif4 as mother wavelet is applied over the raw data. In this approach, detail coefficients of level 3 (D3) and 4 (D4) are used for the calculation of the ratio R due to their equivalence to beta and alpha rhythms, respectively.

In addition, according to our previous results [2], overlapping windows are more appropriate for eye state identification, since they improve system performances and reduce detection times. Therefore, the feature extraction techniques are applied over 10 s windows with an 80% overlap.

For the eye state estimation, SVM and LDA classification algorithms are assessed. With the goal of avoiding bias, each experiment has been repeated ten times, each time implementing a cross-validation process i.e., a different combination of training and test recordings is implemented for each of them.

3. Experimental Results

The performance of each feature extraction technique is analyzed for both eye states. Table 1 shows the mean accuracy obtained for all subjects by the classifiers and the three feature sets.

Table 1. Comparison of the classification accuracies (%). Bold values indicate the best result for each subject.

(a) Accuracy for oE						
	DFT		coif4		JADE+DFT	
Subject	SVM	LDA	SVM	LDA	SVM	LDA
1	99.52	97.71	98.55	94.58	**100.00**	98.31
2	90.96	88.43	81.81	81.20	**93.61**	90.72
3	99.88	95.90	97.47	96.02	**100.00**	93.73
4	**97.95**	89.40	85.30	82.77	85.90	81.33
5	96.63	96.51	70.00	66.51	**97.83**	96.99
6	89.04	70.12	93.25	84.58	**95.54**	73.25
7	**94.70**	83.98	85.30	80.72	93.73	88.31
(b) Accuracy for cE						
	DFT		coif4		JADE+DFT	
Subject	SVM	LDA	SVM	LDA	SVM	LDA
1	**100.00**	**100.00**	98.43	**100.00**	**100.00**	**100.00**
2	87.83	89.88	77.83	82.89	91.45	**92.65**
3	**100.00**	**100.00**	98.67	**100.00**	99.76	**100.00**
4	97.71	**98.92**	86.02	91.81	82.05	88.31
5	96.02	98.43	76.87	79.64	95.54	**97.95**
6	89.40	96.51	94.22	98.19	93.49	**99.52**
7	96.27	**100.00**	84.22	89.76	94.58	99.40

4. Discussion and Conclusions

The presented study in this paper is an extension of our previous work for EEG eye state classification. It can be appreciated from the results that SVM offers the best performance for all subjects and feature sets for oE. Conversely, LDA outperforms SVM for cE. Moreover, it can be seen that DWT presents lower results than the other two techniques. In this regard, JADE combined with DFT seem to provide the highest accuracies for oE using SVM and for cE using LDA. Therefore, it can be concluded that the inclusion of an ICA algorithm for the feature extraction improves the overall system performance. However, a clear decision can not be taken between SVM or LDA, since both methods provide similar results.

Author Contributions: F.L. and F.J.V.-A. have implemented software used in this paper and performed the experiments; F.L. and D.I. have analyzed the data; P.M.C. and A.D. have designed the experiments and head the research.

Funding: This work has been funded by the Xunta de Galicia (ED431G2019/01), the Agencia Estatal de Investigación of Spain (TEC2016-75067-C4-1-R) and ERDF funds of the EU (AEI/FEDER, UE), and the predoctoral Grant No. ED481A-2018/156 (Francisco Laport).

Abbreviations

The following abbreviations are used in this manuscript:

DCT	Discrete Fourier Transform
DWT	Discrete Wavelet Transform
ICA	Independent Component Analysis
LDA	Linear Discriminant Analysis
SVM	Support Vector Machine

References

1. Barry, R.J.; Clarke, A.R.; Johnstone, S.J.; Magee, C.A.; Rushby, J.A. EEG differences between eyes-closed and eyes-open resting conditions. *Clin. Neurophysiol.* **2007**, *118*, 2765–2773.
2. Laport, F.; Dapena, A.; Castro, P.M.; Vazquez-Araujo, F.J.; Iglesia, D. A Prototype of EEG System for IoT. *Int. J. Neural Syst.* **2020**, p. 2050018.
3. Ortiz-Rosario, A.; Adeli, H. Brain-computer interface technologies: from signal to action. *Rev. Neurosci.* **2013**, *24*, 537–552.
4. Cardoso, J.F.; Souloumiac, A. Blind beamforming for non-Gaussian signals. In Proceedings of the F-Radar and Signal Processing, IET, 1993; Volume 140, pp. 362–370.

Proceedings

Acceleration of a Feature Selection Algorithm Using High Performance Computing †

Bieito Beceiro *, Jorge González-Domínguez and Juan Touriño

Computer Architecture Group, CITIC, Universidade da Coruña, 15071 A Coruña, Spain; jgonzalezd@udc.es (J.G.-D.); juan@udc.es (J.T.)

* Correspondence: bieito.beceiro.fernandez@udc.es

† Presented at the 3rd XoveTIC Conference, A Coruña, Spain, 8–9 October 2020.

Published: 1 September 2020

Abstract: Feature selection is a subfield of data analysis that is on reducing the dimensionality of datasets, so that subsequent analyses over them can be performed in affordable execution times while keeping the same results. Joint Mutual Information (JMI) is a highly used feature selection method that removes irrelevant and redundant characteristics. Nevertheless, it has high computational complexity. In this work, we present a multithreaded MPI parallel implementation of JMI to accelerate its execution on distributed memory systems, reaching speedups of up to 198.60 when running on 256 cores, and allowing for the analysis of very large datasets that do not fit in the main memory of a single node.

Keywords: feature selection; mutual information; machine learning; high performance computing; MPI

---

1. Introduction

Feature selection algorithms are data analytics techniques that aim to reduce the size of large datasets by detecting those features that are relevant and discarding the irrelevant ones. These methods have become extremely popular in the last years due to the increasing amount of data that are gathered every day and need to be processed. Moreover, the algorithms used to process data are usually computationally expensive, so the execution times are excessive when datasets grow in size.

In this work, we focus on the Joint Mutual Information (JMI) algorithm [1], since it provides the best tradeoff in terms of accuracy, stability, and flexibility for datasets of various properties, as stated in [2]. JMI follows a filter approach for feature selection, which is, it is used as an independent task performed before the subsequent machine learning algorithm. JMI receives from the user the number of features to select, and it iteratively takes the feature with the highest score, calculated as a tradeoff between its relevance and redundancy. Despite its good results, the main drawback of JMI is its quadratic complexity that prevents its usage to analyze large datasets.

The aim of this work is the development of a new version of the JMI algorithm that is faster than the state-of-the-art solution: the C implementation included in the widely employed and cited suite FEAST [2]. Our implementation must also enable the analysis of huge datasets that do not fit in the memory of conventional computing systems. Both of the goals have been achieved thanks to a novel multithreaded MPI implementation of JMI that can be executed on distributed memory systems, so that it can take advantage of enabling parallel work with several nodes and, thus, reduce the execution time. In addition, this parallel version of JMI is able to divide and distribute extremely large datasets among the memory of all nodes, making it possible to process them.

2. Parallel Implementation of JMI

JMI works with datasets that are represented as matrices, where the rows are the features and the columns are the samples. The output dataset is formatted with a pair for each selected feature. The first element of the pair is its position in the original dataset, while the second one is the score. In this work, we propose a hybrid parallel approach of JMI based on the MPI message-passing library and C++ threads that works with the same input/output format than FEAST and guarantees the same exact results, but with significantly reduced execution times. This has been achieved by using a feature-wise domain decomposition, which is, the features of the input dataset are divided into blocks of the same size so that each block is processed in parallel by a different processing element. Our implementation makes use of this domain decomposition at two levels of parallelism. First, the dataset is distributed among MPI processes. Besides the execution time gains, this MPI parallelization makes it possible the joint usage of the memory of several nodes. Second, the workload of each process is distributed among threads.

Furthermore, two outstanding performance optimizations have been applied to our parallel JMI implementation:

- Semi-distributed data loading to take advantage of the sparse storage format used by the input datasets. While, in a naive data load implementation, one process would read the whole dataset and would scatter it among the other processes, in our semi-distributed approach the root process loads the dataset in sparse format (which fits in memory due to its reduced size), broadcasts it to all of the other processes and then each process transforms the portion of data that needs to work with into a dense matrix.
- Range compression. This is a data preprocessing technique that can reduce the overall execution time and memory usage. It is based on renaming the values of the features so that the range of the values becomes minimal, which has no effect on the results since scores are calculated by value co-occurrences and not by the values themselves. This technique significantly accelerates the creation of the two-dimensional (2D)-histogram that is needed to count co-occurrences between two features. Without range compression, there will be some rows and/or columns that will not affect the result (specifically, those ones created for values that do not appear in the feature), but they are included in the histogram and also have to be processed, increasing the memory consumption and execution time. However, this is avoided when applying our range compression technique, since it assures that only the rows and columns that are needed for the score calculation are included in the histogram. Moreover, this technique introduces almost no overhead due to its linear computational complexity and the possibility of being executed in parallel.

3. Results and Conclusions

The experiments for the analysis of our optimized parallel version of JMI were conducted on 16 nodes of the "Pluton" cluster, a distributed memory system that is based on Intel Xeon processors installed at CITIC. Each node is composed of two processors with eight cores each (16 logical threads using HyperThreading) and 64 GB of main memory. The whole system provides a total of 256 cores (512 logical threads with HyperThreading) and 1 TB of memory. In order to keep the reproducibility of the experiments, we have used five representative datasets that are publicly available in the LibSVM website (https://www.csie.ntu.edu.tw/~cjlin/libsvmtools/datasets/). Their properties are shown at the left half of Table 1.

Table 1 also shows the runtime and speedup (in parentheses) for different number of nodes. Regarding the parallel executions, the same configuration for each node was used, as it proved to be the most efficient one after some preliminary tests: two MPI processes with eight cores each (16 logical threads per MPI process using HyperThreading). However, due to memory problems for the largest datasets (SVHN and E2006), we had to use one MPI process with 16 cores each (32 logical threads) in these cases. Moreover, E2006 is so large (∼512 GB as a matrix in memory) that it does not fit in less than 16 nodes, so it could be executed only using this configuration. As can be seen, the parallel version of JMI performs well and offers excellent scalability (execution times largely decrease when the number of cores increases). Furthermore, it enables the analysis of huge datasets (E2006) that the original version of JMI could not handle due to their size.

Table 1. Dataset properties, execution times (in seconds), and speedups (in parentheses) to select 200 features with JMI.

Dataset	Features	Samples	Original JMI	1 Node	2 Nodes	4 Nodes	8 Nodes	16 Nodes
Epsilon	2000	400,000	1837.59	108.29 (16.97)	57.88 (31.75)	30.86 (59.55)	18.49 (99.38)	12.03 (152.75)
RCV1	47,236	20,242	1834.87	94.86 (19.34)	48.77 (37.62)	27.10 (67.71)	16.38 (112.02)	10.70 (171.48)
News20	62,061	15,935	1813.70	96.38 (18.82)	49.51 (36.63)	27.30 (66.44)	16.12 (112.51)	10.64 (170.46)
SVHN	3072	531,131	8132.46	424.98 (19.14)	264.40 (30.76)	142.95 (56.89)	74.04 (109.84)	40.95 (198.60)
E2006	4,272,227	16,087	–	–	–	–	–	787.64

Author Contributions: Conceptualization, J.G.-D. and J.T.; methodology, B.B., J.G.-D. and J.T.; implementation, B.B.; validation, B.B.; writing—original draft preparation, B.B.; writing—review and editing, J.G.-D. and J.T. All authors have read and agreed to the published version of the manuscript.

Funding: This research was funded by the Ministry of Science and Innovation of Spain (PID2019-104184RB-I00/AEI/10.13039/501100011033), and by Xunta de Galicia and FEDER funds of the EU (Centro de Investigación de Galicia accreditation 2019-2022, ref. ED431G2019/01; Consolidation Program of Competitive Reference Groups, ED431C 2017/04).

Conflicts of Interest: The authors declare no conflict of interest.

References

1. Yang, H.H.; Moody, J. Data Visualization and Feature Selection: New Algorithms for Nongaussian Data. In *Proceedings of the NIPS'99 12th International Conference on Neural Information Processing Systems*; MIT Press: Cambridge, MA, USA, 1999; pp. 687–693.
2. Brown, G.; Pocock, A.; Zhao, M.J.; Luján, M. Conditional Likelihood Maximisation: A Unifying Framework for Information Theoretic Feature Selection. *J. Mach. Learn. Res.* **2012**, *13*, 27–66.

Proceedings

A Doubly Smoothed PD Estimator in Credit Risk†

Rebeca Peláez Suárez [1,*], Ricardo Cao Abad [2] and Juan M. Vilar Fernández [2]

1 Research Group MODES, Department of Mathematics, CITIC, University of A Coruña, 15001 A Coruña, Spain
2 Research Group MODES, Department of Mathematics, CITIC, University of A Coruña and ITMATI, 15782 A Coruña, Spain; ricardo.cao@udc.es (R.C.A.); juan.vilar@udc.es (J.M.V.-F.)
* Correspondence: rebeca.pelaez@udc.es
† Presented at the 3rd XoveTIC Conference, A Coruña, Spain, 8–9 October 2020.

Published: 1 September 2020

Abstract: In this work a doubly smoothed probability of default (PD) estimator is proposed based on a smoothed version of the survival Beran's estimator. The asymptotic properties of both the smoothed survival and PD estimators are proved and their behaviour is analyzed by simulation. The results allow us to conclude that the time variable smoothing reduce the error committed in the PD estimation.

Keywords: probability of default; risk analysis; censored data; survival analysis; nonparametric estimation; kernel estimation

---

1. Introduction

The debts coming from clients with unpaid credits have a important impact in the solvency of banks and other credit institutions. Therefore, one of the most crucial elements that influences the risk in credits is the probability of default (PD). For a fixed time, t, and a horizon time, b, the PD can be defined as the probability that a credit that has been paid until time t, becomes unpaid not later than time $t+b$.

The probability of default conditional on the credit scoring can be written as a transformation of the conditional survival function. Therefore, in Section 2.1 Beran's survival estimator is used to obtain a PD estimator. A time variable smoothing for this estimator is proposed in Section 2.2. In Section 3, both estimators are applied to a real data set. Section 4 contains some concluding remarks.

2. Nonparametric PD Estimators

Let $\{(X_i, Z_i, \delta_i)\}_{i=1}^n$ be a simple random sample of (X, Z, δ) where X is the credit scoring, $Z = \min\{T, C\}$ is the observed maturity, T is the time to default, C is the time until the end of the study or the time until the anticipated cancellation of the credit and $\delta = I_{\{T \leq C\}}$ is the uncensoring indicator. Let x be a fixed value of the covariate X, b a horizon time and $S(t|x)$ the conditional survival function of T. Then, the probability of default in a time horizon $t+b$ from a maturity time t is defined as follows

$$PD(t|x) = P(T \leq t+b | T > t, X = x) = 1 - \frac{S(t+b|x)}{S(t|x)}. \tag{1}$$

Replacing $S(t|x)$ with a nonparametric estimator, $\widehat{S}(t|x)$, in (1), the following estimator for $PD(t|x)$ is obtained:

$$\widehat{PD}(t|x) = 1 - \frac{\widehat{S}(t+b|x)}{\widehat{S}(t|x)}. \tag{2}$$

In [5] the theoretical results that allow to obtain, under general conditions, asymptotic properties for a PD estimator are proved. They are based on these properties for the corresponding estimator of the conditional survival function.

2.1. Beran's Estimator

Beran's survival estimator proposed in [1] is given by

$$\widehat{S}_h^B(t|x) = \prod_{i=1}^{n}\left(1 - \frac{I_{\{Z_i \leq t,\, \delta_i=1\}} w_{i,n}(x)}{1 - \sum_{j=1}^{n} I_{\{Z_j < Z_i\}} w_{n,j}(x)}\right), \tag{3}$$

where the weights are $w_{i,n}(x) = \dfrac{K((x - X_i)/h)}{\sum_{j=1}^{n} K((x - X_j)/h)}$, with $i = 1, \ldots, n$, K is a kernel function and $h = h_n > 0$ is a smoothing parameter. Now, replacing $\widehat{S}(t|x)$ with $\widehat{S}_h^B(t|x)$ in (2), Beran's estimator of the probability of default, $\widehat{PD}_h^B(t|x)$, is available. It was firstly used in [2].

The asymptotic properties of Beran's estimator for the conditional survival function were proven in both [3,4] under certain assumptions. From them, the expressions of the bias and the variance of the estimator $\widehat{PD}_h^B(t|x)$ can be found by using Theorem 1 in [5].

A simulation study was conducted in order to analyse the performance of Beran's estimator. Its behavior was compared with other estimators of the probability of default obtained from estimators of the survival function, including a benchmark method based on proportional hazards models. For more details about the simulation study, see [5].

The results show that the probability of default estimations obtained by means of the estimators built according to (2) are very reasonable, but they have excessive variability and they are very rough curves.

2.2. Smoothed Beran's Estimator

Beran's estimator is smoothed with respect to the covariate, but not with respect to the time variable. This fact along with the survival ratio structure of the PD estimator could be the cause of the instability of the estimations. Therefore, a time variable smoothing of the survival estimator is proposed.

The smoothed Beran's survival estimator is given by

$$\widetilde{S}_{h,g}^B(t|x) = 1 - \sum_{i=1}^{n} s_{(i)} \mathbb{K}\left(\frac{t - Z_{(i)}}{g}\right) \tag{4}$$

where $s_{(i)} = \widehat{S}_h^B(Z_{(i-1)}|x) - \widehat{S}_h^B(Z_{(i)}|x)$ with $Z_{(i)}$ the i-th element of the sorted sample of Z, $\mathbb{K}(t)$ the distribution function of a kernel K and $g = g_n$ is the smoothing parameter for the time variable. Finally, the smoothed Beran's PD estimator, $\widetilde{PD}_{h,g}^B(t|x)$, is obtained by replacing $\widehat{S}(t|x)$ with $\widetilde{S}_{h,g}^B(t|x)$ in Equation (2).

The asymptotic expressions for the bias and the variance of the smoothed Beran's estimator of the survival function have been recently found [6]. The results are too extensive to be shown here. By applying Theorem 1 of [5], the corresponding asymptotic properties of the smoothed Beran's estimator of the PD are obtained.

The simulation study carried out shows that the time variable smoothing significantly reduces the error committed in the PD estimation. This technique implies a considerable increase in the computation time and the improvement is not very noticeable in the estimation of the survival function. However, in the case of the PD, the variability and roughness of the estimations is clearly reduced.

3. Application to Real Data

To illustrate the differences between the estimator based on Beran's and its smoothed version, we obtain the estimation of the conditional survival function and the PD in a real data set. The data consists of a sample of 10,000 consumer credits from a Spanish bank registered between July 2004 and November 2006. The sample contains the credit scoring of each borrower, the observed lifetime and the uncensoring indicator. The sample censoring percentage is 92.8%. The probability of default is estimated using Beran's and smoothed Beran's estimators with $h = 0.05$ and $g = 3$. Figure 1 shows the result.

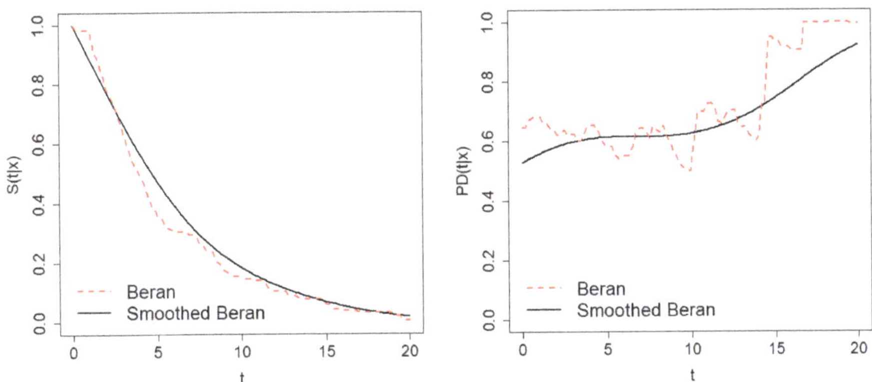

Figure 1. Estimation of $S(t|x)$ (**left**) and $PD(t|x)$ (**right**) at horizon $b = 5$ for $x = 0.8$ by means of Beran's (dashed line) and smoothed Beran's (solid line) estimators on the consumer credits dataset.

4. Conclusions

This work proposes a time variable smoothing for Beran's estimator of the conditional survival function. General asymptotic expressions for the bias and the variance of this estimator are proven. It is used to build a doubly-smoothed PD estimator whose asymptotic properties are also proved. In view of the simulation study carried out, it can be concluded that the smoothed Beran's estimator seems to reduce the estimation error committed when estimating the probability of default.

Work is currently underway to develop a method for choosing the smoothing parameters involved in the above-mentioned estimators. In addition, since the censoring probability is heavy in this context, nonparametric cure models are going to be considered in the study.

Acknowledgments: This research has been supported by MINECO Grant MTM2017-82724-R, and by the Xunta de Galicia (Grupos de Referencia Competitiva ED431C-2016-015 and Centro Singular de Investigación de Galicia ED431G/01), all of them through the ERDF.

References

1. Beran, R. *Nonparametric Regression with Randomly Censored Survival Data*; Technical Report; University of California: Oakland, CA, USA, 1981.
2. Cao, R.; Vilar, J.M.; Devia, A. Modelling consumer credit risk via survival analysis (with discussion). *Stat. Oper. Res. Trans.* **2009**, *33*, 3–30.
3. Dabrowska, D.M. Uniform consistency of the kernel conditional Kaplan-Meier estimate. *Ann. Stat.* **1989**, *17*, 1157–1167.
4. Iglesias-Pérez, M.C.; González-Manteiga, W. Strong representation of a generalized product-limit estimator for truncated and censored data with some applications. *J. Nonparametr. Stat.* **1999**, *10*, 213–244.

5. Peláez Suárez, R.; Cao Abad, R.; Vilar Fernández, J.M. Probability of default estimation in credit risk using a nonparametric approach. *TEST* **2020**, 1–23, doi:10.1007/s11749-020-00723-1.
6. Peláez Suárez, R.; Cao Abad, R.; Vilar Fernández, J.M. Nonparametric estimation of the probability of default with double smoothing. **2020**, under preparation.

Proceedings

Open Source Monitoring System for IT Infrastructures Incorporating IoT-Based Sensors †

Alejandro Mosteiro Vázquez *, Carlos Dafonte and Ángel Gómez

Centro de Investigación en TIC (CITIC)—Department of Computer Science and IT, Universidade da Coruña, Campus de Elviña s/n, 15071 A Coruña, Spain; carlos.dafonte@udc.es (C.D.); angel.gomez@udc.es (Á.G.)

* Correspondence: a.mosteiro@udc.es

† Presented at the III XoveTIC Congress, A Coruña, Spain, 8–9 October 2020.

Published: 3 September 2020

Abstract: This paper introduces the development of a data center monitoring system based on IoT technologies. The system is meant to work as an administrative tool for system administrators in any environment, but mainly focused on data centers, since it integrates sensor and server status data. We are developing a system that gives a broad view of a data center, integrating server data such as CPU and memory usage or network bandwidth with room health parameters such as temperature, humidity, and power consumption or the presence sensors that indicate if there were people inside the room at the time a certain event occurred. As this is a work in progress, in this paper, we present the state-of-the-art of this subject, as well as what we expect to obtain from this project.

Keywords: monitoring; IoT; data center; open source; open hardware

---

1. Introduction

CITIC Research Centre is equipped with a modern DC in a hot-aisle disposition with ten 42U standard racks. All the structural elements in the DC—racks, cooling units, UPS—are from APC. Temperature and humidity sensors are distributed among all the racks, and data from these sensors are collected by a central server capable of sending alerts in case some values deviate from established thresholds. This solution also collects data from the UPS and cooling units, displaying real-time information about load, power consumption, heat dissipation, and cooling demand, amongst others. All of this information is centralized in their proprietary software, StruxuWare Central, and a license is needed for each node in the monitoring infrastructure. With the work described in this paper, we want to create an alternative to the current system using open source and open hardware elements and test if the final product can be as reliable as APC's proprietary system currently used in CITIC's DC and if this new solution can be easily adopted by any DC.

There are several open source monitoring platforms available for IT infrastructure, some of the most used being Zabbix, Nagios, or Icinga, as stated by several review sources such as DNSstuff [1], Opensource.com [2], or iTT Systems [3]. These platforms support similar features: network protocols, operating systems, SNMP. The main difference between them is that the former offers server performance monitoring besides network monitoring out of the box. Adding that Zabbix has a big development community, we decided to use it in this project.

As seen in Figure 1, we propose the installation of Arduino-compatible sensors into racks sending data to a Raspberry Pi with an instance of a home automation platform, such as HomeAssistant or openHAB. Then, we will design a script to communicate all gathered data to the Raspberry Pi from Arduinos to a Zabbix server. This Zabbix server is already deployed and is currently collecting data from a VMware infrastructure, its virtual machines, and several Windows and Linux servers. This way, we will have all relevant data of our DC centralized in our Zabbix server. A web client with responsive and mobile-oriented interfaces will be developed, and it will offer users the possibility to configure alarms, thresholds, notifications, and even interact with some of the elements in the infrastructure.

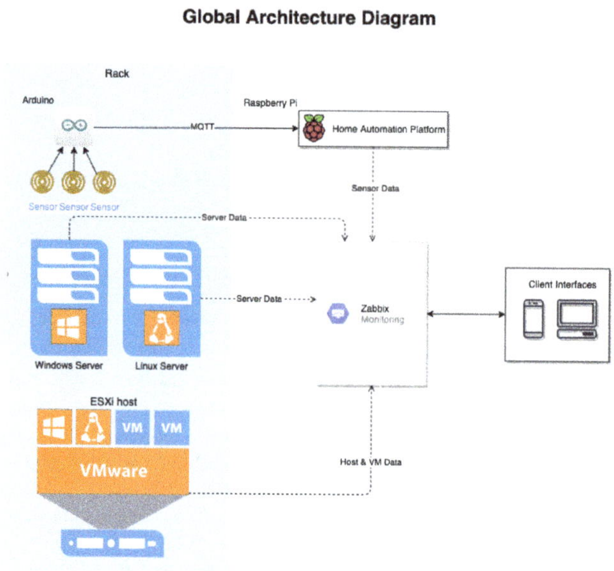

Figure 1. Global Architecture Diagram showing how all components communicate with each other.

We will be using DHT22 sensors for temperature/humidity monitoring and PIR sensors for presence control, all connected to Arduino-compatible boards. Specifically, we thought about using NodeMCU boards as these integrate an ESP8266 chip and allow an easy programming interface. We will also study the possibility of running these boards on battery power and test the battery life expectancy so we can install our sensors even in places where it is impossible to use the main power supplies.

There are some approaches to monitoring sensors with Zabbix like EmonTH sensors Template [4] or the ESP sensors integration for Zabbix [5]. However, these are not a complete generic solution, but templates to monitor specific sensors. Therefore, we will develop our own Zabbix monitoring templates for the chosen hardware keeping in mind that they have to be as generic as possible to monitor all kinds of sensors.

All of the above results in the two main objectives of this project: creating a centralized monitoring tool for CITIC's IT staff and verifying that a system composed of open source and open hardware elements can be as reliable as proprietary alternatives in a real-life working environments.

2. Materials and Methods

One of the main tools we are using in this project is CITIC's virtualization infrastructure. This includes several hosts with VMware's hypervisor, ESXi, managed by VMware's centralization tool, vCenter, where most of the Centre's servers are virtually hosted. All of these servers are physically located in APC racks where sensors are going to be placed. We are also using, as mentioned before, Arduino-compatible sensors, Arduino-based boards and Raspberry Pi boards.

Once our monitoring system is complete, we will collect data from both alternatives: APC StruxuWare Central and CITIC's new monitoring system. Data will be compared, assuming the data from StruxuWare Central are correct and can be established as benchmark data, computing deviations in measurements from our open hardware sensors. If measurements from our system are in acceptable ranges, we will conclude that our system is reliable.

3. Conclusions

In the previous section, we described our monitoring system as an open source monitoring tool that integrates room monitoring with server monitoring using open hardware elements, and there are some articles describing similar approaches such as "Designing an open source maintenance-free Environmental Monitoring Application for Wireless Sensor Networks" [6] or "Environmental monitoring system based on an Open Source Platform and the Internet of Things for a building energy retrofit" [7]. While they offer interesting results for such monitoring applications, these articles cover different types of monitoring. The first one does not include IoT or systems monitoring, and the second one, while more focused on environmental monitoring, is oriented toward building health.

We have experience on the topic of system monitoring, with some environmental sensors on Arduino with Raspberry Pi monitoring tools or DC monitoring on Raspberry Pi with several attached sensors. Therefore, with this project, we will focus on bringing together all knowledge about the topic, integrating data from different sources into one main visualization interface with enhanced features such as real-time alerts or remote control of servers and DC cooling and power elements, offering a novel monitoring system compatible with any DC configuration by applying minor changes to our original design.

Conflicts of Interest: The authors declare no conflict of interest.

Abbreviations

The following abbreviations are used in this manuscript:

IoT	Internet of Things
DC	Data Center

References

1. DNSstuff Review on Open Source Monitoring Tools. Available online: https://www.dnsstuff.com/open-source-network-monitoring-tools (accessed on 30 July 2020).
2. opensource.com Review on Open Source Monitoring Tools. Available online: https://opensource.com/article/19/2/network-monitoring-tools (accessed on 30 July 2020).
3. iTT Systems Review on Open Source Monitoring Tools. Available online: https://www.ittsystems.com/best-open-source-network-monitoring-tools/ (accessed on 30 July 2020).
4. EmonTH Sensors Zabbix Template. Available online: https://share.zabbix.com/scada-iot-energy-home-automation-industrial-monitoring/monitoring-equipment/emonth-sensors (accessed on 30 July 2020).
5. ESP Sensors Integration in Zabbix. Available online: https://github.com/letscontrolit/ESPEasy (accessed on 30 July 2020).
6. Delamo, M.; Felici-Castell, S.; Pérez-Solano, J.J.; Foster, A. Designing an open source maintenance-free Environmental Monitoring Application for Wireless Sensor Networks. Designing an open source maintenance-free Environmental Monitoring Application for Wireless Sensor Networks. *J. Syst. Softw.* **2015**, *103*, 238–247. doi:10.1016/j.jss.2015.02.013.
7. Martín-Garín, A.; Millán-García, J.A.; Baïri, A.; Millán-Medel, J.; Sala-Lizarraga, J.M. Environmental monitoring system based on an Open Source Platform and the Internet of Things for a building energy retrofit. *Autom. Constr.* **2018**, *87*, 201–214. doi:10.1016/j.autcon.2017.12.017.

Proceedings

Fully Automatic Method for the Visual Acuity Estimation Using OCT Angiographies †

Macarena Díaz [1,2,*], Jorge Novo [1,2], Manuel G. Penedo [1,2] and Marcos Ortega [1,2]

1 Centro de investigación CITIC, Universidade da Coruña, 15001 A Coruña, Spain; jnovo@udc.es (J.N.); mgpenedo@udc.es (M.G.P.); mortega@udc.es (M.O.)

2 Grupo VARPA, Instituto de Investigación Biomédica de A Coruña (INIBIC), Universidade da Coruña, 15001 A Coruña, Spain

* Correspondence: macarena.diaz1@udc.es

† Presented at the 3rd XoveTIC Conference, A Coruña, Spain, 8–9 October 2020.

Published: 4 September 2020

Abstract: In this work we propose the automatic estimation of the visual acuity of patients with retinal vein occlusion using Optical Coherence Tomography by Angiography (OCTA) images. To do this, we first extract the most relevant biomarkers in this imaging modality—area of the foveal avascular zone and vascular densities in different regions of the OCTA image. Then, we use a support vector machine to estimate the visual acuity. We obtained a mean absolute error of 0.1713 between the manual visual acuity measurement and the estimated, being considered satisfactory results.

Keywords: OCTA; RVO; VA estimation; biomarkers

1. Introduction

Optical Coherence Tomography by Angiography (OCTA) is the newest imaging modality in the ophthalmic field. This imaging modality is mainly characterized by allowing the visualization of the retinal vasculature non-invasively, being extracted in real time at different levels of depth. The most relevant biomarker in this imaging modality is the Foveal Avascular Zone (FAZ), which represents a region without vascular circulation in the macular area. In our previous work, we have performed the automatic extraction of this biomarker [1]. Visual Acuity (VA) is the representation of a patient's vision. Problems related to the loss of microvasculature lead to a consequent loss of vision. In this work, we present the estimation of the VA of patients with Retinal Vein Occlusion (RVO) using features that were extracted directly from the OCTA images [2,3].

2. Materials and Methods

A dataset containing 860 OCTA images divided into different patient visits is used to validate the proposed method. Each visit consists of four OCTA images representing different depths of the retina (superficial and deep) as well as different zoom levels (3×3 mm^2 and 6×6 mm^2). For a more detailed specification of the used dataset, see Reference [3].

In Figure 1, we can see the steps that were followed to estimate the VA. Using the input OCTA images, we extract the FAZ using the methodology explained in Reference [1] and then we extract the vascular density (VD). To extract the VD, a thresholding process using the Otsu's algorithm was used, followed by a skeletonization procedure. At this point, the image is divided into a grid, typically used in the clinical domain, being the density in each of the regions of this grid calculated. Figure 2 shows a representation of the extractions of these parameters. Finally, using a Support Vector Machine (SVM), we estimate the visual acuity using the FAZ measurements, as well as the densities in each quadrant of the employed grid.

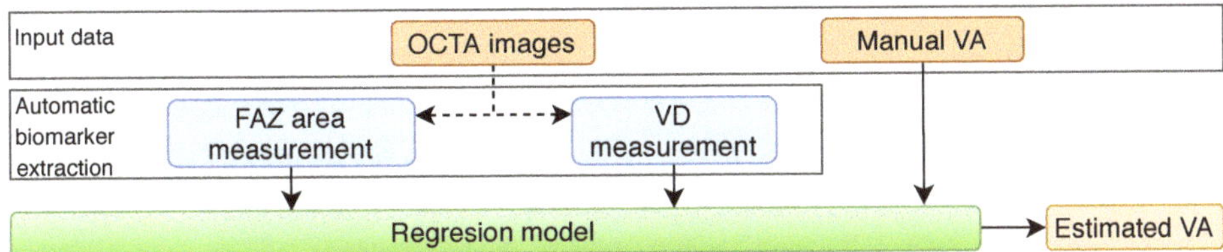

Figure 1. Steps of the presented methodology.

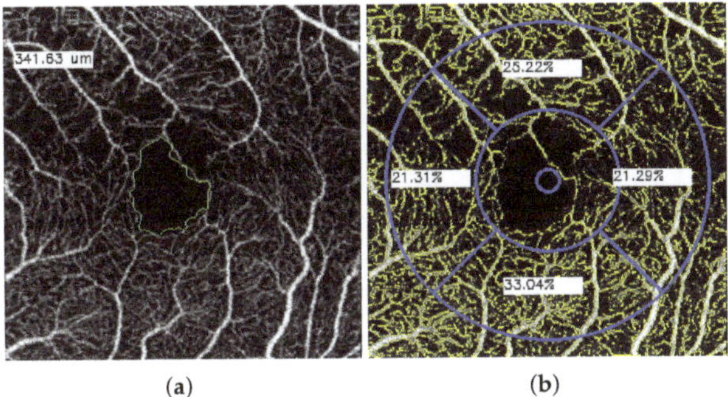

(**a**) (**b**)

Figure 2. Representative examples of the (**a**) FAZ extraction and (**b**) VD measurement process.

3. Results

To validate the methodology we used the Mean Absolute Error (MAE) to evaluate the distance between the manually measured VA and the estimated one with the proposed method. The obtained results reported a MAE of 0.1713 of variation in the VA estimation. Given that the VA takes values in the range [0–1], and the results are obtained in real time (while the manual measurement of the VA is long in time), we can conclude that the results are satisfactory.

Author Contributions: These authors contributed equally to this work. All authors have read and agree to the published version of the manuscript.

Funding: The authors declare no conflict of interest. The funders had no role in the design of the study; in the collection, analyses, or interpretation of data; in the writing of the manuscript, or in the decision to publish the results. This research was funded by CITIC, Centro de Investigación de Galicia ref. ED431G 2019/01 that receives financial support from Consellería de Educación, Universidade e Formación Profesional, Xunta de Galicia, through the ERDF (80%) and Secretaría Xeral de Universidades (20%) and by the Instituto de Salud Carlos III, Government of Spain through the DTS18/00136 research project and by the Ministerio de Ciencia, Innovación y Universidades, Government of Spain through the RTI2018-095894-B-I00 research projects. Also, this work has received financial support from the European Union (European Regional Development Fund - ERDF).

References

1. Díaz, M.; Novo, J.; Cutrín, P.; Gómez-Ulla, F.; Penedo, M.G.; Ortega, M. Automatic segmentation of the foveal avascular zone in ophthalmological OCT-A images. *PLoS ONE* **2019**, *14*, e0212364.
2. Díez-Sotelo, M.; Díaz, M.; Abraldes, M.; Gómez-Ulla, F.; Penedo, M.G.; Ortega, M. A Novel Automatic Method to Estimate Visual Acuity and Analyze the Retinal Vasculature in Retinal Vein Occlusion Using Swept Source Optical Coherence Tomography Angiography. *J. Clin. Med.* **2019**, *8*, 1515.
3. Díaz, M.; Díez-Sotelo, M.; Gómez-Ulla, F.; Novo, J.; Penedo, M.G.; Ortega, M. Automatic Visual Acuity Estimation by Means of Computational Vascularity Biomarkers Using Oct Angiographies. *Sensors* **2019**, *19*, 4732.

Proceedings

Development of Recreational Content with Micro:Bit for Intervention with People with Cerebral Palsy †

Alejandro Lopez-Fernandez [1], Ruben Carneiro-Medin [2], Thais Pousada [3], Betania Groba-González [3] and Adriana Dapena [3,*]

1 Faculty of Informatic, University of A Coruña, Campus de Elviña, 15071 A Coruña, Spain; alejandro.lopez.fernandez@udc.es
2 Aspace Coruña, Sada, 15160 A Coruña, Spain; ruben.cm@aspacecoruna.org
3 CITIC Research Center & University of A Coruña, Campus de Elviña, 15071 A Coruña, Spain; thais.pousada.garcia@udc.es (T.P.); tb.groba@udc.es (B.G.-G.)
* Correspondence: adriana.dapena@udc.es

† Presented at the 3rd XoveTIC Conference, A Coruña, Spain, 8–9 October 2020.

Published: 4 September 2020

Abstract: This paper presents a project carried out to use games in therapies for people with cerebral palsy. A Micro:bit board is used to have a friendly interaction between the user and the game. Through a simple interface, the therapist can manage the parameters of the therapy and see the evolution of the user.

Keywords: cerebral palsy; therapies; Micro:bit; gamification

---

1. Introduction

Cerebral palsy (CP) refers to a group of permanent, non-progressive, developmental disorders that mainly affects movement and posture and with repercussions in daily activity [1]. CP is one of the most common developmental disabilities. Intervention sessions for people with CP are oriented to work on different aspects of daily life in order to improve their personal autonomy and independence using assistive technologies specific to each user's capabilities [2,3]. For that purpose, the therapist defines for each user a set of activities to work on a particular aspect. User motivation is very important to improve the results of therapy. In this sense, gamification is a mechanism that allows to capture the attention of users and with playful performances be working therapeutic aspects [4].

We present a project based on creation an environment that integrates information of the capacity of each user. This allows to characterize their abilities to be able to access playful games based on the work of different physical-cognitive aspects customized for each participant. The control of the games is done through a microcontroller equipped with sensors (accelerometer and compass) and with the possibility of easily connecting to other sensors and actuators compatible with Arduino. Moreover, the tool allows the management of users and records the results of each participation over time allowing the therapist to analyze evolution.

This work is organized as follows. Section 2 presents materials and methods, Section 3 shows the main results obtained using the tool, and Section 4 is devoted to conclusions.

2. Materials and Methods

A Micro:bit board is a tiny computer created by BBC to promote digital creativity (https://microbit.org/). Programs can been created in MakeCoder, Python, or Scratch editor. The program is transferred to Micro:bit to make it run independently of your computer. You can even unplug your micro:bit from the computer, attach a battery pack, and your program still runs. In particular, our project has been developed in Python to integrate the control of micro:bit board with the management of users of the services and monitoring of the results of the therapies.

Three games was created for different purposes:

- Ping-pong. A ball moves from one size to the other size of the screen. Micro:bit allows user to control vertical movement of a bar located in one size of the screen.
- Bank to collect apples. From the top of the screen fall apples that must be trapped in a bowl. Micro:bit allows user to control horizontal movement of the bowl.
- Put balls in a hole. A hole appears in the center of the screen and the user must move the ball across the screen to the hole. Micro:bit allows user to do vertical and horizontal movement .

During a session with an user, the therapist chooses the service using the menu shown in Figure 1a. The therapist chooses the game from the games associated to this service (see Figure 1b) and determines the parameters: time, sensitivity, objective, etc. To obtain good results during intervention, it is very important that these parameters be adapted to user and, for this reason, the tool allows you to stop a game at any time so that you can adjust them. Once the game is chosen, the user plays using micro:bit. For example, Figure 1c shows a game "bank to collect apples" that allows user to make movements on the horizontal axis. After the game is over, the therapist can see the results of that game and also access the results of previous games (see Figure 1d).

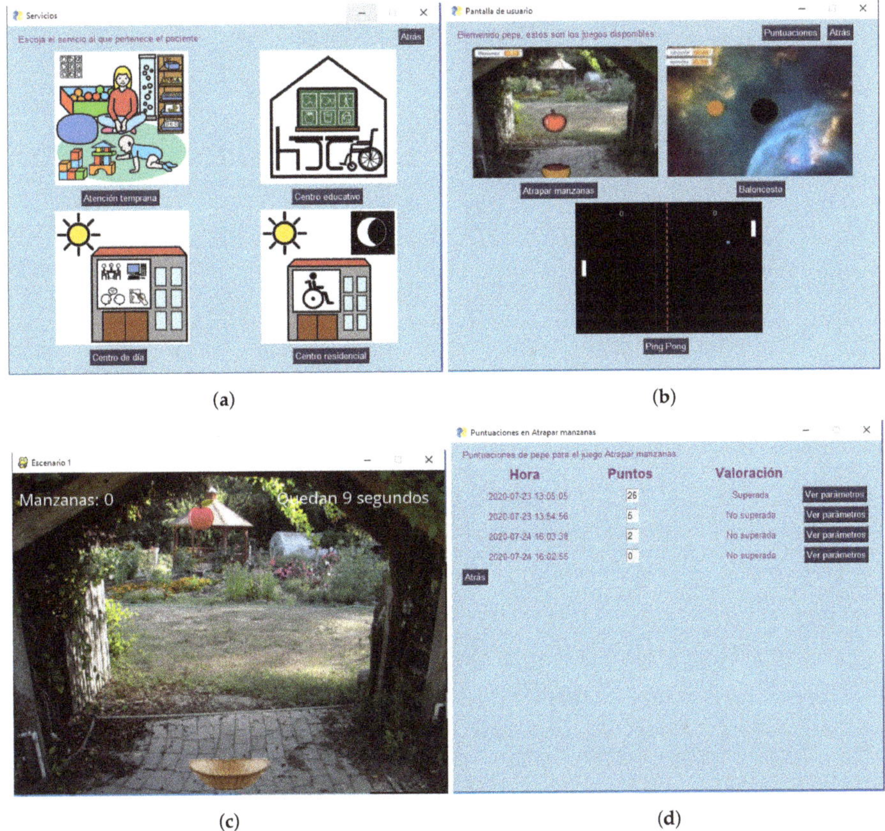

(a) (b) (c) (d)

Figure 1. Developed tool: (**a**) Selection of services, (**b**) selection of games, (c) example of a game, and (**d**) user's results.

3. Results

With the aim of verifying the usefulness of using the tool with real users, a test was carried out at Aspace in A Coruña. Aspace's mission is to improve the quality of life of people with cerebral palsy and other related disabilities by defending their rights and supporting families, services to associated entities, and institutional cooperation.

Prior to the session, the therapist familiarizes the users with the service. First, the therapist chose the service, the user for this service, and the game with the default parameters. Table 1 resumes the results obtained during the intervention with two users. As can be seen, the parameters have been adjusted during the session. After finishing the session, both the users and the therapists shown their satisfaction with the use of the tool. Therapists highlight the ease of setting up game parameters and keeping track of user results.

Table 1. Resume from the sessions with users.

User	Games	Micro:bit	Parameters	Rating	User Comments
User 1	Bank to collect apples.	In a cap	Sensibility: 3; Time: 30 s; Speed (apple): 3; Speed (bowl): 3	6	Too fast; I need more time
User 1	Bank to collect apples.	In a cap	Sensibility: 3; Time: 60 s; Speed (apple): 2; Speed (bowl): 2	8	I like it
User 1	Put balls in a hole.	In a cap	Sensibility: 3; Time: 60 s; Speed (ball): 2	1	Too fast
User 1	Put balls in a hole.	In a cap	Sensibility: 1; Time: 60 s; Speed (ball): 2	6	I like it
User 1	Ping-pong	In a cap	Sensibility: 3; Time: 60 s; Speed (bar): 1	6	I can't see the bar
User 2	Bank to collect apples.	In a swing table	Sensibility: 2; Time: 60 s; Speed (apple): 1; Speed (bowl): 1	4	I like it
User 2	Bank to collect apples.	In a cap	Sensibility: 2; Time: 60 s; Speed (apple): 1; Speed (bowl): 1	8	I like it

4. Discussion and Conclusions

The development of this project has allowed creating a useful tool for the therapy of people with cerebral palsy. The use of Python has allowed the development of a single environment for the control of games using micro:bit and for monitoring therapies. Carrying out some tests with the users has made it possible to refine some important aspects for their future use by the Aspace entity.

Author Contributions: A.L.-F. implemented software; R.C.-M. performed the interventions; T.P., B.G.-G. and A.L.-F. analyzed the results; A.L.-F. and A.D. wrote the paper and A.D. heads the research.

Funding: Centro de Investigación de Galicia CITIC and Campus Innova (agreement I+D+ 2019-20) is funded by Consellería de Educación, Universidade e Formación Profesional from Xunta de Galicia and European Union (European Regional Development Fund - FEDER Galicia 2014-2020 Program) by grant ED431G 2019/01 and Universidade da Coruña. Partially supported by the Spanish Ministry of Science (Challenges of Society 2019) PID2019-104323RB-C33.

Conflicts of Interest: The authors declare no conflicts of interest.

References

1. Bax, M.; Goldstein, M.; Rosenbaum, P.; Paneth, A.; Paneth, N.; Dan, B.; Jacobsson, B.; Damiano, D. Proposed definition and classification of cerebral palsy. *Dev. Med. Child Neurol.* **2005**, *47*, 571—576.
2. Pousada, T.; Pareira, J.; Groba-González, B.; Nieto, L.; Pazos, A. Assessing mouse alternatives to access to computer: A case study of a user with cerebral palsy. *Assist. Technol.* **2014**, *26*, 33–44.

3. Pousada, T.; Pereira, J.; Groba-González, B.; Nieto-Rivero, L. Assistive technology based on client-centered for occupational performance in neuromuscular conditions. *Medicine* **2019**, *98*, e15983.
4. Lopes, S.; Magalhaes, P.; Pereira, A.; Martins, J.; Magalhaes, C.; Chaleta, E.; Rosario, P. Games Used with Serious Purposes: A Systematic Review of Interventions in Patients With Cerebral Palsy. *Front. Psychol.* **2019**, *9*, 1712. doi:10.3389/fpsyg.2018.01712.

Proceedings

Predicting Gastric Cancer Molecular Subtypes from Gene Expression Data †

Marta Moreno [1,2,*], Abel Sousa [3,4,5,6], Marta Melé [7], Rui Oliveira [2,8] and Pedro G Ferreira [1,2,3,4,*]

1 Department of Computer Science, Faculty of Sciences, University of Porto, 4169-007 Porto, Portugal
2 University of Minho and INESC TEC, 4200-465 Porto, Portugal; rui.oliveira@inesctec.pt
3 Ipatimup—Institute of Molecular Pathology and Immunology of the University of Porto, 4200-465 Porto, Portugal; abels@ipatimup.pt
4 i3s—Instituto de Investigação e Inovação em Saúde da Universidade do Porto, 4200-135 Porto, Portugal
5 Graduate Program in Areas of Basic and Applied Biology, Abel Salazar Biomedical Sciences Institute, University of Porto, 4050-313 Porto, Portugal
6 European Molecular Biology Laboratory, European Bioinformatics Institute, Wellcome Genome Campus, Cambridge CB10 1SD, UK
7 Life Sciences Department, Barcelona Supercomputing Center, Barcelona, 08034 Catalonia, Spain; marta.mele@bsc.es
8 Department of Informatics, University of Minho, 4710-057 Braga, Portugal
* Correspondence: up201305416@fc.up.pt (M.M.); pgferreira@fc.up.pt (P.G.F.)
† Presented at the 3rd XoveTIC Conference, A Coruña, Spain, 8–9 October 2020.

Published: 7 September 2020

Abstract: Stomach cancer is a complex disease and one of the leading causes of cancer mortality in the world. With the view to improve patient diagnosis and prognosis, it has been stratified into four molecular subtypes. In this work, we compare the results of multiple machine learning algorithms for the prediction of stomach cancer molecular subtypes from gene expression data. Moreover, we show the importance of decorrelating clinical and technical covariates.

Keywords: gene expression; gastric cancer; disease classification; machine learning

---

1. Introduction

Several large-scale projects, such as TCGA (The Cancer Genome Atlas) or ICGC (International Cancer Genome Consortium), have studied dozens of tumor types through the analysis of hundreds of samples with several molecular assays of the genome, epigenome, proteome, transcriptome and the respective clinical data. One such example is stomach adenocarcinoma (STAD), representing nearly 5% of new cancer cases worldwide [1]. STAD is a complex disease, with a mortality rate almost equivalent to its incidence.

The molecular profiling of more than four hundred tumor cells with five different assays has allowed for the identification of four novel STAD sub-types with different diagnostic and prognostic value [2]. However, extensive characterization of tumor samples is not always possible due to clinical, technical or budget limitations.

Previous studies have shown that strong outcome predictor signatures can be derived from RNA data in cancer [3]. These studies indicate that gene expression carries sufficient signal for the accurate prediction of phenotypes. For this reason, we believe that the genetic alterations observed in different STAD molecular subtypes should be reflected in differential tissue gene expression

Here, we set to investigate if it is to possible to develop a predictive tool that, based on transcriptome profiling with RNA-seq, can predict stomach cancer samples according to the proposed stratification. In order to minimize the effect of possible unwanted sources of variation in the data, we have analyzed the impact of pre-processing the data, taking into account the effect of the available covariate information.

2. Materials and Methods

STAD-specific transcriptome data were obtained from the TCGA Research Network (https://www.cancer.gov/tcga). Samples with insufficient clinical information were excluded. As features, only coding genes with a median Fragments per Kilobase per Million (FPKM) value higher than 1 were retained (Figure 1) and their values were log2 transformed.

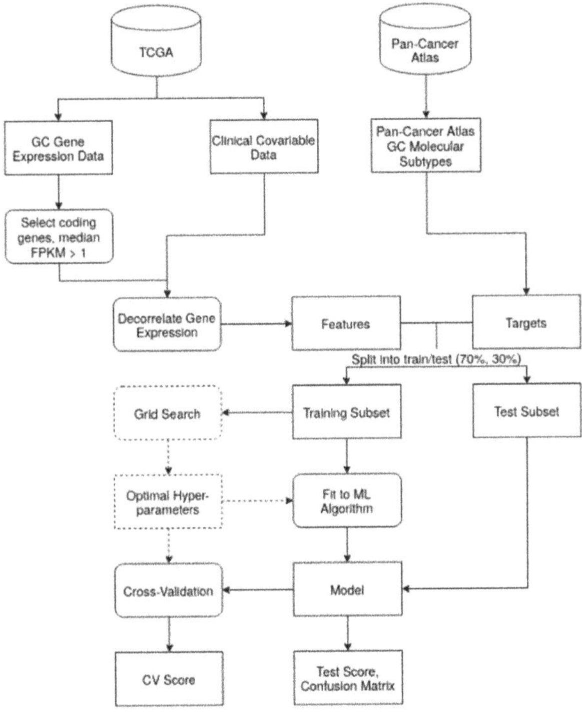

Figure 1. Diagram of the pipelines used in this work. Steps with a dashed line were only performed on pipelines with hyper-parameter optimization.

Technical or clinical factors may correlate with both the features and the target STAD molecular subtypes, possibly confounding machine learning (ML) predictions. Without a decorrelation step, the model may thus over- or under-estimate the effect of the features on the target variable. As a data pre-processing step, we regressed out the possible confounding effects of the covariates on the gene expression data through a multiple linear model:

$$g_i = \beta_0 + \beta_1 age + \beta_2 gender + \beta_3 race + \beta_4 age\_diagnosis + \beta_5 distant\_metastasis + \beta_6 primary\_tumor + \beta_7 icd\text{-}10 + \beta_8 morphology + \beta_9 diagnosis + \beta_{10} prior\_malignancy + \beta_{11} tissue + \beta_{12} tumor\_stage +$$

where g_i represents the gene expression for gene i, β_0 is the intercept, β_i i ∈ (1, ..., 12) is the regression coefficients for the covariates, and is the noise term.

The residuals of the model, obtained as the difference between the real gene expression value (gi) and the predicted expression ($\hat{g}i$), were used as the expression phenotype.

After this step, several ML pipelines were devised with the goal of predicting STAD molecular subtypes from RNA-seq data (chromosomal instability (CIN) 61.45%, Epstein–Barr virus (EBV) 7.54%, genomically stable (GS) 12.85%, microsatellite instability (MSI) 18.16%; see Figure 2a). First, the dataset was split into stratified training (n = 250) and test (n = 108) sets. Each algorithm learned from the training set's features to build prediction models, with or without hyper-parameter optimization. Cross-validation was performed to test the model's performance on sampled portions of the training data, with subsequent validation using the unseen test set.

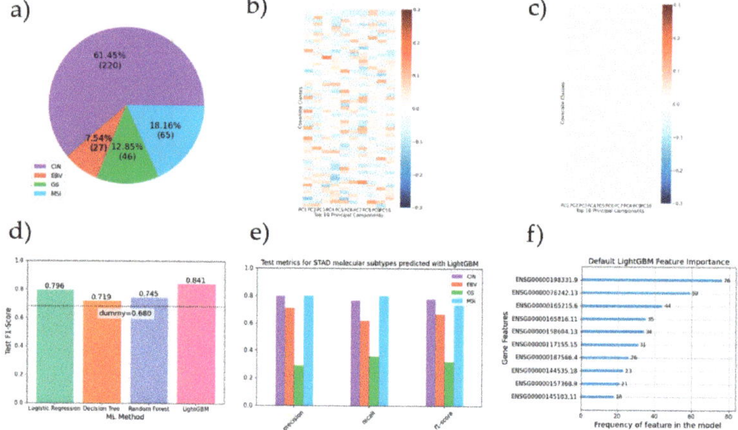

Figure 2. (**a**) Distribution of STAD molecular subtypes in the data (N = 358). (**b**,**c**) Correlation heatmap between clinical covariates and the top 10 gene expression principal components. (**b**) Before covariate decorrelation. (**c**) After covariate decorrelation. (**d**) Distribution of test f1-scores across methods, as compared to the dummy estimator's (which always predict most frequent class) score. (**e**) Test metrics obtained using LightGBM with default settings, stratified by class. (**f**) The top 10 gene features by their importance for the LightGBM default model.

3. Results

Several covariates possessed significant correlation (ranging from -0.14 to 0.17) with the top 10 principal components for gene expression (Figure 2b). As expected, all covariate correlation was lost after gene-wide covariate decorrelation (Figure 2c).

Despite a heavy class imbalance (Figure 2a), all machine learning models outperformed a dummy estimator that always predicted the most frequent class, with an average 8% improvement across methods (Figure 2d). There were also notable differences in performance between algorithms, with the best performer, LightGBM, having a test F1-score 5.6% better than the second best, logistic regression. By contrast, there was no significant difference between results of models using default algorithm hyper-parameters and those obtained following hyper-parameter optimization. On a per class basis, the CIN sub-type exhibits the best results (Figure 2e). The top 10 most informative gene features for the best performing model (LightGBM default) are shown in Figure 2f. Of special interest, the second most contributing gene, ENSG00000076242 (MLH1), is a tumor suppressor gene whose epigenetic silencing is associated to MSI tumors.

4. Discussion

Machine learning methods show promise for the prediction of molecular subtypes in STAD, with even the simplest methods performing better than random chance. However, perhaps due to the small sample size and/or imbalance of the data, hyper-parameter optimization offered no performance improvements.

Funding: This work was supported by the FCT (Fundação para a Ciência e a Tecnologia) research grant Ph.D. Studentship SFRH/BD/145707/2019 and the research grant IF/01127/2014, funded in the scope of the FCT Investigator Exploratory Project: "Understanding the impact of acquired and germline genetic variants in the complexity of gastric cancer"; GenomePT project (reference 22184): "National Laboratory for Genome Sequencing and Analysis"; QREN L3 project (reference NORTE-01-0145-FEDER-000029): "Mapping genetic and phenotypic heterogeneity in HER2 positive cancers to anticipate and counteract resistance phenotypes".

Conflicts of Interest: The authors declare no conflict of interest.

References

1. World Health Organization; International Agency for Research on Cancer (IARC). *GLOBOCAN 2018: Estimated Cancer Incidence, Mortality and Prevalence Worldwide in 2018*; WHO: Geneva, Switzerland; IARC: Lyon, France, 2018. Available online: https://gco.iarc.fr/today/online-analysis-pie (accessed on 23 July 2020).
2. Cancer Genome Atlas Research Network. Comprehensive molecular characterization of gastric adenocarcinoma. *Nature* **2014**, *513*, 202–209, doi:10.1038/nature13480.
3. Byron, S.A.; Van Keuren-Jensen, K.R.; Engelthaler, D.M.; Carpten, J.D.; Craig, D.W. Translating RNA sequencing into clinical diagnostics: Opportunities and challenges. *Nat. Rev. Genet.* **2016**, *17*, 257–271, doi:10.1038/nrg.2016.10.

Proceedings

Electronic Health Records Exploitation Using Artificial Intelligence Techniques †

Carla Guerra Tort [1,*], Vanessa Aguiar Pulido [2], Victoria Suárez Ulloa [3], Francisco Docampo Boedo [4], José Manuel López Gestal [4] and Javier Pereira Loureiro [1]

1 CITIC-Research Center of Information and Communication Technologies, University of A Coruña, 15071 A Coruña, Spain; javier.pereira@udc.es
2 Department of Computer Science, University of Miami, Coral Gables, FL 33146, USA; vanessa@cs.miami.edu
3 Institute for Biomedical Research of A Coruña (INIBIC)-Fundación Profesor Novoa Santos, 15006 A Coruña, Spain; Victoria.Suarez.Ulloa@sergas.es
4 Instituto Médico Quirúrgico San Rafael, 15009 A Coruña, Spain; fdocampo@imqsanrafael.es (F.D.B.); jlopez@imqsanrafael.es (J.M.L.G.)
* Correspondence: c.gtort@udc.es
† Presented at the 3rd XoveTIC Conference, A Coruña, Spain, 8–9 October 2020.

Published: 9 September 2020

Abstract: The exploitation of electronic health records (EHRs) has multiple utilities, from predictive tasks and clinical decision support to pattern recognition. Artificial Intelligence (AI) allows to extract knowledge from EHR data in a practical way. In this study, we aim to construct a Machine Learning model from EHR data to make predictions about patients. Specifically, we will focus our analysis on patients suffering from respiratory problems. Then, we will try to predict whether those patients will have a relapse in less than 6, 12 or 18 months. The main objective is to identify the characteristics that seem to increase the relapse risk. At the same time, we propose an exploratory analysis in search of hidden patterns among data. These patterns will help us to classify patients according to their specific conditions for some clinical variables.

Keywords: electronic health record (EHR); Artificial Intelligence (AI); relapse; respiratory diseases

---

1. Introduction

The electronic health record (EHR) represents the digital version of a patient's medical history. In an EHR system, data is stored in a collection of tables where each record corresponds to a patient's healthcare episode. EHRs constitute a rich source of information, including demographic data (age, gender, address, ...), administrative data and a wide range of clinical information (clinical notes, diagnoses, procedure-treatments, lab test, medical imaging...) [1–3]. The knowledge extracted from EHRs can be used in clinical decision support, epidemiological and predictive tasks, population care improvement and pattern recognition [2,4]. For this reason, the exploitation of EHRs has aroused interest of researchers in the last years [5,6]. Nevertheless, EHRs have some characteristics that make this goal hard to achieve. Heterogeneity, noise, incompleteness, redundancy or the inconsistent representation of data are some of the challenges to cope with. In this context, exploratory analysis and preprocessing steps play a fundamental role [7,8].

Artificial Intelligence (AI) has become a key tool for EHR exploitation. Machine Learning and Deep Learning have been successfully used to identify new risk factors, patterns and medical associations [6,9]. In addition, recent studies show the potential of these modern techniques to make predictions better than the traditional existing methods [9–11].

In this project, we propose the use of AI to exploit and extract value from EHR data. More concretely, we focus our study on the analysis of relapse rates in patients suffering from the most prevalent diagnoses in our data set. We consider as a relapse the return of a disease time after its apparent overcoming. We will construct a Machine Learning model to predict whether a patient will have a recurrence in less than 6, 12 or 18 months (depending on diagnosis). This model will allow us to identify the characteristics that seem to increase the relapse risk in those patients. At the same time, we will carry out exploratory analysis in search of hidden patterns among data. We hope the results help us to classify patients according to their specific conditions.

2. Data Set Description

Anonymous patient data were extracted from the San Rafael Hospital database. Records range from January 2000 to January 2020. Main diagnoses and procedures are encoded in both ICD-9 and ICD-10, so the data is divided in two codification sets. ICD-9 set consists of 156,362 records and 89,211 patients. ICD-10 set consists of 32,069 records and 25,013 patients. More information about the sets is given in Table 1. Demographic and clinical features acts as predictive variables.

Table 1. Numeric description of ICD-9 and ICD-10 sets.

	Records	Patients	Diagnoses	Procedures
ICD-9	156,362	89,211	4147	1581
ICD-10	32,069	25,013	2691	2555

3. Present Work

Currently, the study is in the preprocessing phase. The most frequent diagnoses have been identified by a descriptive study of the data set. Table 2 shows these main diagnoses and the associated recounts. Among all the most prevalent diagnoses, we have selected those related to respiratory problems. We discarded the diagnoses of traumatology and varicose veins because they were not considered relevant to this specific research.

After selecting the ICD-9 and ICD-10 codes of interest, the sets will be unified in order to procure a larger and completed data set. Null and missing values will be removed to ensure data quality. In addition, the Machine Learning models will be defined. Once we obtain a clean data set, the next steps will allow us to recognize the most explanatory predictive variables for the chosen diagnoses. For this task, we will apply a Principal Component Analysis (PCA) [12].

Table 2. Most prevalent diagnoses in the data set.

	Records	Patients	Relapses
Dorsopathies	10,177	8228	1250
Varicose veins	9700	7981	1568
Arthropathies	9437	8137	1116
Respiratory infections	7722	4349	1466
Rheumatism	6601	5943	534

Author Contributions: All the authors have contributed to the conceptualization of the paper and the design of the research; methodology and AI models definition, V.A.P. and V.S.U.; data acquisition, F.D.B. and J.M.L.G.; writing—original draft preparation, C.G.T.; writing—review and editing, J.P.; All authors have read and agreed to the published version of the manuscript.

Funding: Centro de Investigación de Galicia CITIC is funded by Consellería de Educación, Universidades e Formación Profesional from Xunta de Galicia and European Union (European Regional Development Fund—FEDER Galicia 2014-2020 Program) by grant ED431G 2019/01. Partially supported by the Spanish Ministry of Science (Challenges of Society 2019) PID2019-104323RB-C33.

Conflicts of Interest: The authors declare no conflict of interest.

References

1. Pham, T.; Tran, T.; Phung, D.; Venkatesh, S. Predicting healthcare trajectories from medical records: A deep learning approach. *J. Biomed. Inform.* **2017**, *69*, 218–229, doi:10.1016/j.jbi.2017.04.001.
2. Yadav, P.; Steinbach, M.; Kumar, V.; Simon, G. Mining Electronic Health Records (EHRs): A Survey. *ACM Comput. Surv.* **2018**, *50*, 85:1–85:40, doi:10.1145/3127881.
3. Martínez-Romero, M., Vázquez-Naya, J. M., Pereira, J., Pereira, M., Pazos, A., & Baños, G. The iOSC3 system: Using ontologies and SWRL rules for intelligent supervision and care of patients with acute cardiac disorders. *Comput. Math. Methods. Med.* **2013**, doi:10.1155/2013/650671.
4. Shickel, B.; Tighe, P.J.; Bihorac, A.; Rashidi, P. Deep EHR: A Survey of Recent Advances in Deep Learning Techniques for Electronic Health Record (EHR) Analysis. *IEEE J. Biomed. Health Inform.* **2018**, *22*, 1589–1604, doi:10.1109/JBHI.2017.2767063.
5. Marier, A.; Olsho, L.E.W.; Rhodes, W.; Spector W.D. Improving prediction of fall risk among nursing home residents using electronic medical records. *J. Am. Med. Inform. Assoc.* **2016**, *23*, 276–282, doi:10.1093/JAMIA.
6. Panahiazar, M.; Taslimitehrani, V.; Pereira, N.; Pathak, J. Using EHRs and Machine Learning for Heart Failure Survival Analysis. *Stud. Health Technol. Inform.* **2015**, *216*, 40–44, doi:10.3233/978-1-61499-564-7-40.
7. Yue, L.; Dongyuan, T.; Weitong, C.; Xuming, H.; Minghao, Y. Deep learning for heterogeneous medical data analysis. *World Wide Web* **2020**, 1–23, doi:10.1007/s11280-019-00764-z.
8. Miotto, R.; Li, L.; Kidd, B.A.; Dudley, J.T. Deep Patient: An Unsupervised Representation to Predict the Future of Patients from the Electronic Health Records. *Sci. Rep.* **2016**, *6*, 26094, doi:10.1038/srep26094.
9. Weng, S.F.; Reps, J.; Kai, J.; Garibaldi, J.M.; Qureshi, N. Can machine-learning improve cardiovascular risk prediction using routine clinical data? *PLoS ONE* **2017**, *12*, e0174944, doi:10.1371/journal.pone.0174944.
10. Kawaler, E.; Cobian, A.; Pessig, P.; Cross, D.; Yale, S.; Craven M. Learning to Predict Post-Hospitalization VTE Risk from EHR Data. In Proceedings of the 12th AMIA Annual Symposium, Chicago, Illinois, USA, 3–7 November 2012; pp. 436–445.
11. Wong, N.C.; Lam, C.; Patterson, L.; Shayegan, B. Use of machine learning to predict early biochemical recurrence after robot-assisted prostatectomy. *BJU Int.* **2019**, *123*, 51–57, doi:10.1111/bju.14477.
12. Maćkiewicz, A.; Ratajczak, W. Principal components analysis (PCA). *Comput. Geosci.* **1993**, *19*, 303–342, doi:10.1016/0098-3004(93)90090-R.

Proceedings

Machine Learning to Compute Implied Volatility from European/American Options Considering Dividend Yield †

Shuaiqiang Liu [1], Álvaro Leitao [2,*], Anastasia Borovykh [3] and Cornelis W. Oosterlee [4]

1 Delft University of Technology, 2628 Delft, The Netherlands; s.liu-4@tudelft.nl
2 CITIC, University of A Coruña, 15071 A Coruña, Spain
3 Imperial College London, London SW7 2AZ, UK; a.borovykh@imperial.ac.uk
4 Centrum Wiskunde & Informatica, 1098 XG Amsterdam, The Netherlands; c.w.oosterlee@cwi.nl
* Correspondence: alvaro.leitao@udc.gal
† Presented at the 3rd XoveTIC Conference, A Coruña, Spain, 8–9 October 2020.

Published: 15 September 2020

Abstract: Computing implied volatility from observed option prices is a frequent and challenging task in finance, even more in the presence of dividends. In this work, we employ a data-driven machine learning approach to determine the Black–Scholes implied volatility, including European-style and American-style options. The inverse function of the pricing model is approximated by an artificial neural network, which decouples the offline (training) and online (prediction) phases and eliminates the need for an iterative process to solve a minimization problem. Meanwhile, two challenging issues are tackled to improve accuracy and robustness, i.e., steep gradients of the volatility with respect to the option price and irregular early-exercise domains for American options. It is shown that deep neural networks can be used as an efficient numerical technique to compute implied volatility from European/American options. An extended version of this work can be found in [1].

Keywords: implied volatility; neural networks; dividend yield; European options; American options

---

1. Problem Formulation

The Black–Scholes model for pricing European options reads,

$$\frac{\partial V_{eu}}{\partial t}+\frac{1}{2}\sigma^2S^2\frac{\partial^2 V_{eu}}{\partial S^2}+(r-q)S\frac{\partial V_{eu}}{\partial S}-rV_{eu}=0, \tag{1}$$

where r and q are the risk-less interest rate and continuous dividend yield, respectively.

For American options, the original Black–Scholes equation becomes a variational inequality,

$$\frac{\partial V_{am}}{\partial t}+\frac{1}{2}\sigma^2S^2\frac{\partial^2 V_{am}}{\partial S^2}+(r-q)S\frac{\partial V_{am}}{\partial S}-rV_{am}\leq 0, \tag{2}$$

where the free boundary condition is $V_{am}(S,t)\geq H(K,S_t)$, and the terminal condition is $V_{am}(S,T)=H(K,S_T)$. We can employ a numerical method, here the COS method [2], to solve the American pricing model. In this work, we will focus on put options. The European/American Black–Scholes solution is denoted by $V_{eu/am}=BS_{eu/am}(\sigma,S_0,K,\tau,r,q,\alpha)$.

Given an observed market option price V^{mkt} (European or American), the Black–Scholes implied volatility σ^* is defined by

$$BS(\sigma^*;S_0,K,\tau,r,q,\alpha)=V^{mkt}. \tag{3}$$

There does not exist a closed-form expression of the inverse function for neither American-style or European-style options. A popular way is to formulate the above problem into a minimization problem,

$$\min_{\sigma^* \in R^+} BS(\sigma^*; S_0, K, \tau, r, q, \alpha) - V^{mkt} \tag{4}$$

There are several root-finding numerical algorithms to solve (4), for example, Newton–Raphson,

$$\sigma^*_{k+1} = \sigma^*_k - \frac{V(\sigma^*_k) - V^{mkt}}{BS'(\sigma^*_k)} = \sigma^*_k - \frac{V(\sigma^*_k) - V^{mkt}}{\text{Vega}(\sigma^*_k)}, \; k = 0, \ldots. \tag{5}$$

However, some issues are likely to arise when using derivative-based algorithms, see Figure 1.

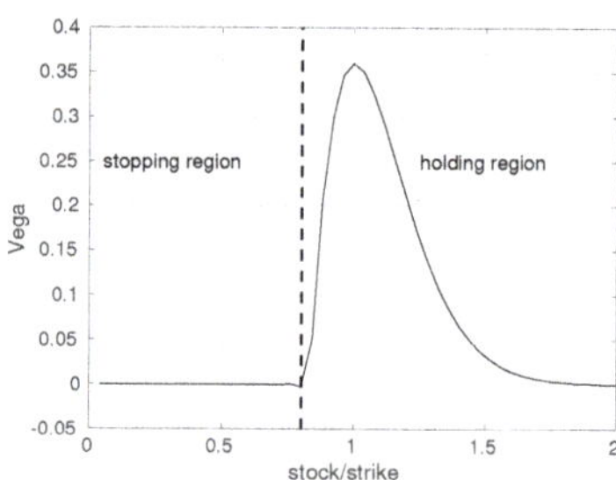

Figure 1. The Vega for American options: One tail is near zero, and the other tail is flat (zero value).

2. Methodology

In order to avoid an iterative algorithm, we provide a data-driven approach for directly approximating the inverse function of (3) via neural networks. Mathematically, an artificial neural network (ANN) can be represented as a composite function,

$$\mathrm{F}(\mathbf{x}|\boldsymbol{\Theta}) = f^{(L)}(\ldots f^{(2)}(f^{(1)}(\mathbf{x}; \boldsymbol{\theta}_1); \boldsymbol{\theta}_2); \ldots \boldsymbol{\theta}_L), \tag{6}$$

where $\mathbf{x}$ stands for the input variables, $\boldsymbol{\Theta}$ for the hidden parameters (i.e., weights and biases), L for the number of hidden layers.

The implied volatility defined by Equation (3) can be written as an inverse function of the pricing model,

$$\sigma^* = BS^{-1}(V^{mkt}; S, K, \tau, r, q, \alpha), \tag{7}$$

where $BS^{-1}(\cdot)$ denotes the inverse Black–Scholes function (European-style or American-style). Please note that the definition domain of (7) is the continuation region Ω_h for American-style options. We use a deep neural network to approximate the inverse Black–Scholes function,

$$\sigma^* = BS^{-1}(V^{mkt}; S, K, \tau, r, q, \alpha) \approx \mathrm{NN}(V^{mkt}; S, K, \tau, r, q, \alpha). \tag{8}$$

Thus we do not need any iterative algorithm to solve (3).

2.1. ANN for European Implied Volatility

The inverse Black–Scholes function probably gives rise to steep gradients of the volatility with respect to the option price, especially for deep OTM/ITM options. It is known that the ANN has difficulties accurately representing such gradients. Here we employ the gradient-squashing technique, as in [3], to address this issue,

$$\hat{V}^P_{eu} = \log\left(V^P_{eu}(S,t) - \max(Ke^{-r\tau} - S_t e^{-q\tau}, 0)\right). \tag{9}$$

2.2. ANN for American Implied Volatility

For training the ANN to compute implied volatility from American options, there are two steps, due to the early-exercise feature. First, we need to compute again the gradient-squashed time value of an American option,

$$\hat{V}_{am}^{P} = \log\left(V_{am}^{P}(S_t, t) - \max(K - S_t, Ke^{-r\tau} - S_t e^{-q\tau}, 0)\right). \quad (10)$$

Second, the effective definition domain Ω_h of the inverse function (8) is numerically found based on the generated samples.

3. Numerical Results

In addition to being robust, the neural network (IV-ANN) solver is much faster than an iterative numerical solver to compute implied volatility from European/American options including dividends, see Tables 1–4.

Table 1. Artificial neural network (ANN) hyper-parameters.

Parameters	Values
Hidden layers	4
Neurons(each layer)	400
Activation	ReLU
Initialization	Glorot
Optimizer	Adam
Batch size	1024

Table 2. Model parameter ranges.

	Parameters	Range
Inputs	Stock price (K/S_0)	[0.3, 1.8]
	Time to maturity (τ)	[0.08, 2.5]
	Risk-free rate (r)	[0.0, 0.25]
	Dividend yield (q)	[0.0, 0.25]
	Scaled time value ($\hat{V}$)	-
Output	Volatility (σ)	(0.01, 1.05)

Table 3. Performance of IV-ANN.

Phase	European Options				American Options			
	MSE	MAE	MAPE	R^2	MSE	MAE	MAPE	R^2
Training	1.72×10^{-7}	3.17×10^{-4}	6.99×10^{-4}	0.9999976	7.12×10^{-7}	5.66×10^{-4}	1.42×10^{-4}	0.999990
Testing	1.94×10^{-7}	3.35×10^{-4}	7.39×10^{-4}	0.9999972	1.93×10^{-6}	6.52×10^{-4}	2.35×10^{-3}	0.999974

Table 4. Computational cost based on 20,000 option prices.

Method	GPU (s)	CPU (s)	Robustness
Newton-Raphson	19.68	23.06	No
Brent	52.08	60.67	Yes
Bi-section	337.94	390.91	Yes
IV-ANN	0.20	1.90	Yes

References

1. Liu, S.; Leitao, A.; Borovykh, A.; Oosterlee, C.W. On Calibration Neural Networks for extracting implied information from American options. *arXiv* **2020**, arXiv:2001.11786.
2. Fang, F.; Oosterlee, C.W. Pricing early-exercise and discrete barrier options by Fourier-cosine series expansions. *Numer. Math.* **2009**, *114*, 27, doi:10.1007/s00211-009-0252-4.
3. Liu, S.; Oosterlee, C.W.; Bohte, S.M. Pricing Options and Computing Implied Volatilities using Neural Networks. *Risks* **2019**, *7*, 16, doi:10.3390/risks7010016.

MDPI
St. Alban-Anlage 66
4052 Basel
Switzerland
Tel. +41 61 683 77 34
Fax +41 61 302 89 18
www.mdpi.com

Proceedings Editorial Office
E-mail: proceedings@mdpi.com
www.mdpi.com/journal/proceedings

www.ingramcontent.com/pod-product-compliance
Lightning Source LLC
LaVergne TN
LVHW071028170726
843515LV00006B/1233